# NEW PLYWOOD BOATS

## *And a Few Others*

*By the same author*

Boats to Go

Low-Resistance Boats

Multihull Voyaging

# *NEW PLYWOOD* **BOATS**

*And a Few Others*

**Thomas Firth Jones**

SHERIDAN HOUSE

First published 2001 by
Sheridan House Inc.
145 Palisade Street
Dobbs Ferry, NY 10522



*Library of Congress Cataloging-in-Publication Data*
Jones, Thomas Firth, 1934-
New plywood boats : and a few others / Thomas Firth Jones.
p. cm.
Includes index.
ISBN 1-57409-096-8 (alk. paper)
1. Boatbuilding. 2. Plywood boats. 3. Boats and boating. I. Title.

VM321 .J68 2001
623.8'4--dc21

2001020560

Edited by Catherine Degnon
Production management by Quantum Publishing Services, Inc., Bellingham, Washington
Designed by Jill S. Mathews
Photos by the author unless otherwise indicated

Printed in the United States of America

ISBN 1-57409-096-8

# Contents



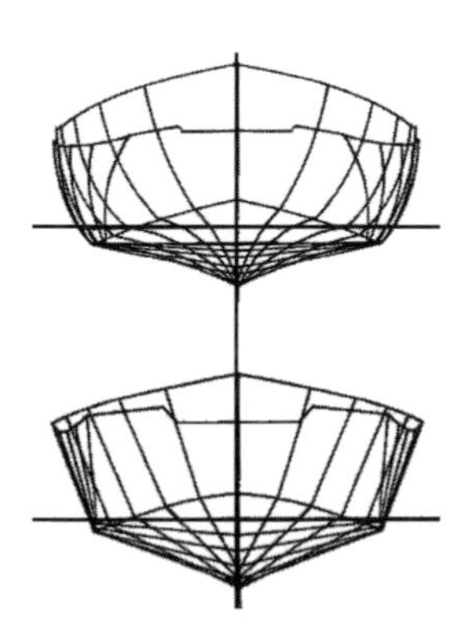

# Introduction

This book describes boats I've built or designed in the last ten years, some for customers and some for my own pleasure. Looking back over them, I am surprised by how many are built of plywood, although the techniques vary. Several others are solid or cored fiberglass, several are clinker-planked cedar, and one is a modern high-tech version of the old canvas-over-stringers canoe. Some of these boats are muscle powered, and some are wind or motor powered. My discussion touches on both their designs and their building.

Weight is the greatest single factor in boat design—and perhaps in boat use as well. A large automobile may be more luxurious to ride in than a small one, and undoubtedly it displays to better advantage its owner's ability to pay for it, but inevitably a small car is more fun to drive. So it is with a boat, despite the fact that in most conditions boatspeed is dependent on length. Boats are slow anyway compared to cars, because water is so much denser than air; and the sensation of speed is what is fun, not the absolute speed. None of the boats in this book is over 30 feet long (that is also the case in my first boat book, initially called *Low-Resistance Boats* but titled *Boats to Go* in its latest edition, which described earlier designing and building projects and gained me many friends).

The physical laws governing the performance of single-hulled boats have been well understood for more than a century now, ever since the British naval architect and engineer William Froude separated the two causes of resistance: friction and wave-making. The laws governing multihull sailboats are pretty well understood too, after half a century of experiment and development. But the building of boats is ever new, with new materials and techniques coming along all the time and altering the weight of some boats so radically from that of their predecessors that they must be redesigned.

Plywood compromises boat design because it will take certain shapes but not others. With several boats in this book, I have been unwilling to put up with plywood's limitations and have built some part of a mostly plywood boat of another material—for example, strip planking or fiberglass in one manner or another. Some amateur builders are keen to cover wooden boats with fiberglass but leery of using fiberglass alone, because they think glass has to have a mold. In truth, a well-made wooden boat seldom benefits from glass sheathing, and a fiberglass shape can often be made quickly and easily with a rudimentary form. However, fiberglass is tedious and unpleasant to finish.

Plywood comes with a factory-made finish. Properly used, it is very light for its strength and stiffness, and it is cheap. Boat-quality plywood costs only about \$1 per square foot for each $^{1}/_{8}$ inch of thickness. Marine plywood is less expensive than all other good boatbuilding materials except steel and ferrocement, but some people want to save even more money by using plywood intended for housebuilding (or perhaps dog-house building). To them I say, good luck!

Yet even the misguided folks who want to build a boat of underlayment are not motivated by economy alone. After all, used boats are even cheaper! A friend of ours who wanted a boat trailer stopped at a house where a good-looking one was for sale, with a decent-looking fiberglass runabout strapped down to it. The sign read $150 for both. Our friend knocked on the door and told the owner, "I'll give you a hundred and fifty dollars for the trailer alone." The owner replied, "I had in mind to sell the package."

Reader, even if you are just perusing this introduction, boats are more to you than mere appliances. Together we belong to a club, along with the older guys that my wife, Carol, and I sometimes see when we're coming down New York's East River on the way home from a cruise to New England. They are hanging over the rail at the United Nations Plaza, studying the boat traffic through binoculars. For them, every line and every detail really does matter. Here is a chance for you to study one boat that they have seen—ours—as well as a score of others.

This book contains study plans for every one and complete plans for six. In order of building difficulty, they are *Chingadera, Speed King, Dandy* dinghy, Weekender, Melonseed, and Buy-boat. Some first-time builders have completed boats larger and more complex than any of these, but others have poured years into relatively simple designs and abandoned them unfinished. As you would (or should) before entering into any marriage, before starting to build a boat, you need to think honestly of your prospective spouse's patience, skills, and stamina as well as your own. Despite what people say, there are no friendly divorces.

## Acknowledgments

For enlightenment on various subjects sprinkled throughout this book, I thank Susanne Altenburger, Phil Bolger, Vance Buhler, Dave Carnell, Milt Edelman, Tom LaMers, Curt Read, and John Yank. Thanks also go to Bob Hicks of *Messing about in Boats,* Keith Lawrence of *Boatbuilder,* and Mike O'Brien of *WoodenBoat.* For help with every phase of the writing and revising, I extend profound thanks to my wife, Carol.

Several chapters of this book first appeared in different form in *Boatbuilder* magazine.

# 1

# Row and Paddle Boats

"Anything that goes by itself!" Jim Wolf, an old guy working at Yank Boats when I started there twenty years ago, was talking about tools, not boats, but the comment applies to both. If a sail will push a boat fairly reliably and a motor somewhat more reliably (and faster!), who in his right mind would push one with his own muscular energy? Not John Yank, for sure. He and his current girlfriend own a canoe with an electric outboard. Recently they tooled up the Tuckahoe River to Head of River, and he told me that they got there and halfway back at full throttle before the battery went flat. Then they had to paddle not just the canoe and themselves but also the battery and motor back home. And after that they had to recharge the battery, perhaps carrying it up to the house. Whether they saved effort by not making the whole trip under paddle I'm not sure. Never mind; for a good part of the way, the boat did go by itself.

However, there are places a muscle-powered boat can go that other boats just can't. Here in the Pine Barrens of South Jersey we have hundreds of miles of small "rivers"—most of them creeks not even 15 feet wide—that wind under bridges, over fallen trees, and around turns that would have even the shortest runabout backing and filling. Canoe liveries do enough business here to create traffic jams on summer weekends. The only other boats you see on these "rivers" are kayaks, and they work less well because their paddles are clumsy in small waters and they can't be steered from the ends: they don't have bow and stern thrusters as canoes, in effect, do.

In addition, muscle-powered boats are quickest and most efficient for genuinely short voyages, such as from a mooring to a dock. An engine must be started and does not invariably do so. A sailing rig must be set up. Oars or paddles weigh less than motors or sailing rigs and so are the best choice when a boat must be carried some distance overland before being used in water. But none of these factors explains the vast popularity that muscle-powered boats now enjoy.

Kayaks are currently the most fashionable of the muscle-powered boats, just as they were a hundred years ago. (Between then and now, rowboats and canoes were the favorites.) Exercise is also fashionable nowadays, to the point where some people use rowing machines to have the exercise without the nuisance of getting out on the water (one neighbor even works out on his rowing machine every morning facing his bedroom window, which looks out on the Tuckahoe River).

However, many people who know how to sail and know how to run a motorboat, who are not looking for exercise and are not slaves to fashion, still prefer muscle-powered boats. Designer Phil Bolger is one such fan, and he thinks that in the thirty years he has owned his kayak *Kotick,* he may have put 5,000 miles on her. He says that on the water he doesn't like to concentrate as much as one has to in a sailboat and ought to in a motorboat; he likes being able to let his mind wander.

My wife, Carol, is another fan. Her kayak, the *Kingston Wailer,* measures 11 feet, 10 inches by 23 inches and weighs just 25 pounds with a cold-molded arc bottom and virtually plumb topsides. We can't guess how far it's taken her in the nineteen years she's owned it, but more often than not when the two of us are going out on the river in small boats, I take a sailboat and she takes the *Wailer*. She is a small woman, but she can carry it down to the dock and launch it herself, which she really can't do with even the lightest sailboat. In addition, she can't swim, and she never felt comfortable in small boats until she was persuaded that she didn't look funny wearing a lifejacket. She says she feels secure in her kayak in whatever condition of wind and water may arise. She likes the ease of beaching and of wriggling into the marsh grass to salvage a Spackle bucket, a milk crate, a fender brought up by the tide.

Even I, who usually prefer to sail, sometimes take a kayak out alone of an afternoon. Perhaps the wind is too light (contrary to what people say, there always is *some* wind) for a sailboat to cope with the river tide. Perhaps it's an unusually warm day in winter, when the water is too cold to risk even the faintest possibility of capsize. Perhaps I need to be home again at a certain time and can't risk the wind going light. A kayak is so quiet; you can get closer to wildlife before they're startled than you can in other craft. The boat is so simple, and you can set your own pace. That is why I built the first boat in this chapter for myself.

# Chingadera

*Dread Bob,* 2 feet longer and 3 inches wider than the *Wailer* but built in the same manner, I had for fifteen years and never much liked. Nothing pleasing about the *Wailer* quite came off in the larger version, and eventually I was sick of using it and even of looking at it. I cut it up and made a nice little bonfire, although 30 pounds of thin wood doesn't burn very long. Perhaps it could have been sold, but selling a boat is work—and much less pleasant work than boatbuilding; and as it cost only

$75 to build and at that point needed paint, the work of selling it might not have yielded minimum wages. Besides, there was some satisfaction in saying to the hulk, "I never liked you. Take that!"

As the plans show, *Chingadera* is flat bottomed and double chined. I wanted a more stable boat than *Bob*. Many times every year when the tide is high and the wind is west, we cartop the kayaks to Head of River, where the Tuckahoe is only a Pine Barrens creek, and paddle lazily home, 4 miles in about an hour and a half, with wind and current doing much of the work. The launching site is a steep, mossy bank below a highway bridge. I hold the *Wailer* for Carol but must then get into my own boat unaided. *Bob* never did flip in this maneuver but came close nearly every time.

*Chingadera* has only an inch more waterline beam than *Bob* but feels very much more stable. The flat bottom really does help and so do the upper chines, which are exactly at waterline for half the boat's length. This shape is really not too different from what is found in mass-produced canoes, which are likely to be flat-bottomed athwartships for nearly three-quarters of their width and have no rocker at all until very near the ends. It's a good shape for inexperienced or careless boaters. Mass-produced kayaks—they come in astonishingly short lengths—usually have a great deal of rocker and tend to wiggle-waggle through the water. My older son bought a fiberglass 13-footer in Australia and had no satisfaction at all from it until he glued on a skeg. The *Wailer* and *Bob* had skegs from the start.

It seems that there is no perfect kayak shape for every condition of wind and water, and for that reason my younger son in Italy owns half a dozen kayaks. A

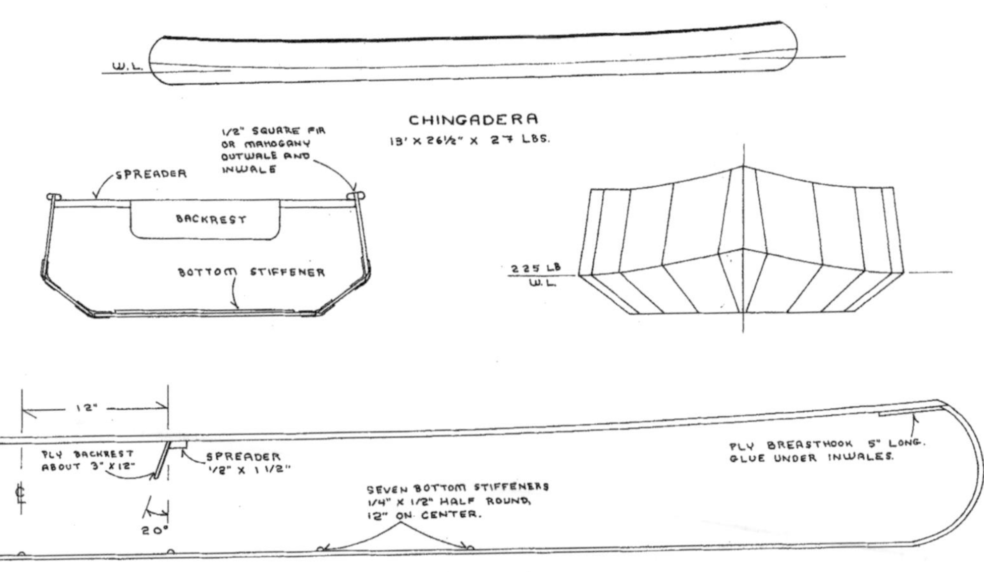


Chingadera

Carol in *Chingadera* at Head of River

bottom with no rocker—and most designs from the legendary builder J. Henry Rushton have virtually none—produces the straightest running boat in light wind or with a strong wind forward. With a strong wind aft the rockerless hull with its deep forefoot tends to slew around broadside, sometimes more than can be corrected by paddling exclusively on one side, and the only way to control the boat is to drag a paddle blade. Once I built a clinker cedar kayak to Rushton's lines, and it slews as much as *Chingadera*, while Carol's *Wailer*, which has no forefoot and a skeg aft, is relatively easy to keep on course downwind.

The stability of a flat bottom is good, but wetted surface is also a concern. At the speed a kayak goes—4 knots is really a spurt in a 13-footer and 2 $^1/_2$ knots is probably a paddling average—wetted surface is the main source of resistance, and wave-making is secondary. A kayak of the same displacement and beam as *Chingadera* but with a flat bottom athwartships would have 11 percent more wetted surface and would require at least 11 percent more energy to push through the water. Exercise, no doubt, is a fine thing, and it's strange that exercisers seek out sleek boats and speedy bicycles when they could get more exercise like Sisyphus by pushing a rock up a hill. But for me, being out on the water is the fun, and the more moderate the exercise, the better I like it.

Another dimension that has a great effect on wetted surface is length, and this is trickier. Using the same two sheets of plywood, *Chingadera* could be lengthened 3 feet, which would increase the theoretical hull speed by $^1/_2$ knot (1.34 times the

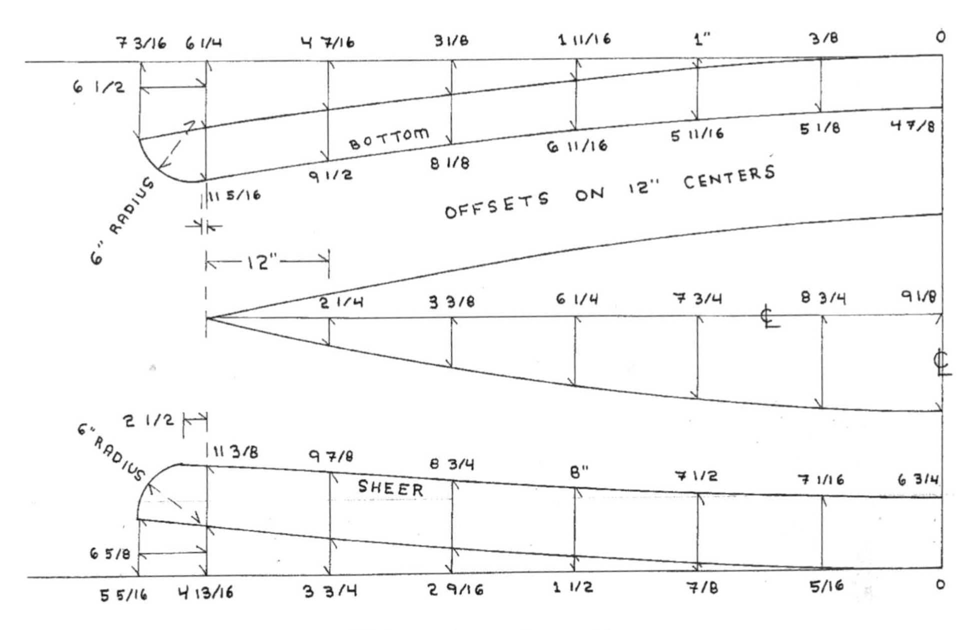


*Chingadera* ply cutting

square root of the waterline). Length-to-beam ratio would increase from 6:1 to 7.4:1, which would diminish wave-making resistance a bit. Probably lengthening the boat would be a good idea for a really heavy and muscular paddler: As designed, *Chingadera* displaces 225 pounds, which is meant to float boat, paddle, paddler, and gear, and at 16 feet she could displace 50 pounds more. Wetted surface would increase by nearly 20 percent, however.

It's beside the point to talk about hull speed in a kayak. Even racers don't maintain it on courses more than a mile long. Their boats are nearly 20 feet long with not much more than half the waterline beam of *Chingadera,* so they probably have less wetted surface along with far less stability. Although racers are stronger than you and me, they are still using their arm muscles to propel the boat, and arms are feeble compared to legs, let alone to sails or motors.

Too short a muscle-powered boat does not carry its way between strokes. That can be observed with the beamy, roto-molded kayaks commonly sold in camping stores and even more obviously with short, fat prams—because rowing requires a longer pause between strokes. Every stroke is almost like starting out again from a dead stop, and before long he or she is thinking that a motor—just a little motor—wouldn't be a bad idea. Twelve feet may be nearly the minimum length that a paddling boat should be if the owner expects to enjoy using it.

All our kayaks are undecked; some people label such boats "double-paddle canoes," although to me the difference between a kayak and a canoe lies in where you sit and in the number of blades the paddle has. Sitting on a seat or kneeling against a spreader as the Indians did and swinging a single-bladed paddle makes the vessel a canoe. The center of gravity is higher in either case, so canoes must be proportionately beamier than kayaks.

The purpose of a kayak deck is to keep out water, most often in waves or rapids. White-water canoes often have decks too, with kayaklike holes for the paddlers to slither into. I'm not sure a deck adds much weight because freeboard can be reduced, which also reduces windage. But a deck makes a kayak less habitable, harder to clean, and less suitable for toting things, whether they be food or clothing or salvage picked up during the trip. In addition, if you carry a second paddle in a decked boat (Carol often uses a canoe paddle in the tight upper reaches of the Tuckahoe and breaks out the double paddle when she gets to less constricted water), it has to be lashed on deck. And in still water a deck is annoying, and it complicates the building. Nearly all the kayaks we see on the river are decked because they are store-bought, and that's what's available.

The tumblehome in *Chingadera*'s topsides precludes your barking your hands on the gunnels while paddling. It could have been more severe, but I like to set a small spritsail (30 square feet is plenty) when the wind is fair and shift my weight to windward when need be. A decked kayak has very low freeboard at the sheer, which makes it pleasant to paddle, but there is no good place to rest the paddle when drifting, which we do at least half the time. Some people swing a kayak paddle so that it comes nearly vertical on every stroke; then the height and width of gunnel matter less. We keep our paddles more nearly horizontal because it is less work and brings fewer drips into the boat.

*Chingadera* is built of 5-millimeter okoume plywood. This is a lovely panel, five plies thick and a weight of just over $^1/_2$ pound per square foot. I wouldn't recommend anything thinner. I did once build a racing kayak for Carol of 3-millimeter okoume. The boat was 16 feet long and partially decked and weighed only 19 pounds. But the method was tortured ply, so the compound curves added stiffness, and *Chingadera*'s panels are flat or moderately curved in one direction. The racer's plywood proved very vulnerable to local impact, so the boat was not at all suited to careless use. Even a tree branch floating in the river was a real hazard.

People do love to build boats like this one of nonmarine plywood. Lauan underlayment is a favorite because it is so inexpensive, and admittedly the *Wailer*'s topsides are underlayment and are still fine after all these years. But the boat has never been left outdoors, even for one night, and is underlayment as good now as it was nineteen years ago? Certainly Douglas fir marine plywood, a perfectly useful product twenty, and even ten, years ago, is now fit only for subfloors both according to John Guzzwell, author of *Modern Wood Yacht Construction,* and to my own observation. And, of course, subfloors are now made of chipboard. Capitalism is forever driving down the quality of goods. With underlayment you can be sure that logs that would once have been rejected by the lumber mill no longer are; the glue is a little thinner; and the making of the panels is speeded up and less closely supervised.

If we want to build boats that are as good as we used to, we must use materials with a higher-quality label than we did in the past. At this writing the cost of the panels of 5-millimeter okoume for *Chingadera* is $120, and there is a considerable amount left over to start another project. The whole boat, including epoxy, glass tape, a little lumber, and paint, should cost less than $200. Underlayment would not save half that amount and no labor at all, and if it started to delaminate, as we see so often happen...

Let us begin building. My *Chingadera* was made on a strongback over molds, but most people like to build kayaks without strongback or molds, so that's what the plans show. The panels are not drawn to make maximum use of the plywood sheet, but rather to be easy for you to lay out on your own sheet. You can see that with careful cutting it is possible to use the intermediate and topsides panels as patterns to trace another one of each on this single sheet of plywood; so only the other half of the bottom need come out of the second sheet. With a great deal of scarfing, it might be possible to get the other half-bottom out of a single sheet, but I doubt doing so would be a good use of your time.

A single scarf in each panel is recommended, which means that half the pieces you cut need a 2-inch extension. I make scarfs in thin plywood by running an electric plane across the face grain and then finishing with a 7-inch disc grinder with 60-grit hardback. If you don't have these tools, scarfs can be made with a hand plane, but the danger with scarfs always is that their surfaces wind up convex, making a lumpy join, so keep a straight-edge handy to check them. If you don't want to scarf, you can use ply butt blocks about 4 inches wide, which adds about 1 pound to boat weight. Builder Harold H. "Dynamite" Payson and Dave Carnell discovered at about the same time that plywood panels can be butt-joined with a piece of glass tape on each side. I have not tried this method, but Dave says it's as strong as a scarf, which undoubtedly means that he has tested it. He's a retired chemist, and he loves to test things.

Wire ties about 6 inches on center are sufficient except at stem and stern, where each pair of panels should have three ties. I use steel baling wire, not copper wire, because on the inside I tape only between ties, not over them, and when the epoxy resin has kicked off I take the wires out. Interstices can be filled with epoxy putty, and the boat is strong enough if the outside taping is continuous. At stem and stern, glassing between the wire ties is too troublesome, so there it's best to use the conventional copper wire. House wiring—the single-strand stuff that runs inside the walls—suffices.

Eight-ounce glass is best, in strips about 3 inches wide, but it's better to cut your own from cloth than to buy ready-made tape because the selvage edge of purchased glass tape is lumpy to sand. Some builders say that glass tape is best cut at 45 degrees to the weave, because then both warp and woof are holding the panels together; but 45-degree tape is awfully stretchy, with twice as many ends to ravel and wick up. Tape is best put on early in the day so that a second layer of resin can be applied when the first one is tacky, to save sanding twice.

When *Chingadera*'s wire ties are all in place and snugged up, the hull should be turned upright and set on a flat surface. Because her bottom is to be flat fore and aft as well as athwartships, you may need to put a few braces down from the ceiling. She needs a 24-inch temporary spreader amidships from gunnel to gunnel, and two more spreaders 3 feet forward and aft. These spreaders are best tacked or screwed in place and left there until both outside and inside taping are completed. After that you have a nice mess of sanding to do, with dust everywhere and the surface never becoming truly fair. To avoid sanding and the inevitable lumpy result of tape-seaming, I usually prefer to build boats with beveled wooden stringers; but in *Chingadera* taped seams were used to save weight. My boat weighs just 27 pounds, only 2 pounds more than Carol's *Wailer*, which is in every dimension a smaller boat.

Sam Devlin, who champions taped seaming in his tortured ply designs and probably has as much experience as anyone with the method, tells me that he cannot get a fair surface after taping unless he puts on an immense amount of filler and sands it off. Rather than do that, he uses a change of color or a molding to distract the eye from a lumpy join: in other words, smoke and mirrors. Dynamite Payson, who builds prototypes of many of the Bolger designs that he sells, is likely to cover the whole boat with a layer of fiberglass, which of course makes the lumps less obvious but adds incredibly to the labor of building and the weight of the finished product.

When *Chingadera* is taped and sanded, the few pieces of lumber (Douglas fir or Honduras mahogany) can be put in: the inwales and outwales and the permanent spreader with its backrest. The sole stiffeners can be troublesome because they are small for metal fasteners, but they definitely are needed because flat 5-millimeter okoume cannot span 18 inches and bear the weight of the paddler stepping into the boat. With the stiffeners the bottom comes close to having the strength of 1/2-inch plywood. The little sticks should begin just inside the tape and be positioned with a 3/8-inch tack at each end. Then a few scrap two-by-fours spanning at least two stiffeners each can weigh them down and allow most of the epoxy that oozes out from under them to be cleaned up with a putty knife while it is still soft.

If you want a sailing rig for down- and cross-wind work (trying to make canoes and kayaks sail upwind usually requires a hiking board, and tacking these craft is a big job), now's the time for step and partners. My step is a 1/2-inch plywood donut, and the partners are 1/2 inch by 2 1/2 inches to take a mast of 1 1/2-inch fir closet rod. The step is 2 feet, 4 inches from the bow. For the times when I want to paddle hard, there's a 1/2-inch-by-1-inch toe brace, but its distance from the backrest varies with your dimensions; the easiest way to get it right is to sit in the boat and have a friend clamp it in as you direct.

A taped-seam boat definitely does not want high-gloss paint. I prefer latex semigloss house paint for most boats; I find that it lasts about as well as one-pot polyurethane paint and a great deal better than alkyd paint. That stuff never was as good as latex, and since the law requiring the lead be taken out of it went into effect, it's become very much not-as-good. Alkyd paint doesn't abrade or stain as much as latex, but it chalks almost at once in sunlight, and it likes to blister or peel, while latex just sloughs away, a little bit every year. I use latex primer, too. People may tell you that alkyd primer "soaks in" better, probably because it takes longer to dry. However, paint is a surface coating, and nothing that is put onto sound wood soaks in significantly unless it is applied with pressure, like creosote or the green arsenic compound now used on yellow pine.

A half-used can of latex paint stays good for years and needs only a light stirring when next opened, although if the can is opened too often the paint can rust through the can's lip, because one of its ingredients is water. Polyurethane forms a skin, which Dave Carnell attributes to the oxygen in the air. To exclude air from an opened paint or varnish can, Dave suggests unscrewing the tip from a propane torch, removing the jet, and squirting some gas into the can. Propane is heavier than air and so forces it up and out. Then the can lid can be cautiously put on. This really does work, though doing it repeatedly will suck in the sides of the can till it looks like the head of an aged cat.

*Chingadera* has been a success for three years now. When our friend Franny Read comes down the river in her kayak with us, she is at some pains to paddle as slow as we do; but Franny likes a workout, and her boat is a sleek, tortured-ply 19-footer, and she always wins the women's class at the annual Tuckahoe River kayak race. We are not embarrassed. We are out on the river just to be there and to see something of nature, and we enjoy Franny's company when she pauses to let us catch up.

I am more comfortable in *Chingadera* than I was in the dreadful *Dread Bob.* I can move around in her more and get into her more easily, and I believe she goes a little farther with every stroke I take. She weighs a few pounds less, which means very little on the water, but when I carry her up to the boat shed or lift her off a roof rack, every single ounce matters—for some people, in fact, weight may be a factor in how often the boat is used.

Like boats in many Latin countries, our kayaks always have eyes on the bows to "see where they are going," although the skipper's eyesight is helpful, too. You may be wondering about the name of this one. It is Spanish, and it means "love token," roughly translated. Carol and I were watching a video of a 1940 John Ford western, and the cowboy heroes were driving a herd north from Mexico. One showed another a gold bauble he was bringing home to his girlfriend, and the other said, "That's a real *chingadera!*" We looked at each other. Did he really say that? So we rewound the tape and played it again, and sure enough, he did say that. The censors must not have understood Spanish, and if you don't either, you had best look up the verb *chingar* yourself.

## Geodesic Airolite Canoe

The canoe is *the* American boat. The kayak is also North American in origin, but from a part of the continent that few of us know about except from books and that even fewer of us ever aspire to visit. It was developed by a people whose language and customs are so different from ours that no boat of European, African, or Asian origin could seem more alien to us. But not long ago, we think, canoes were everywhere here! Few of us can view a body of water without imagining it a few centuries ago, full of Indians in their canoes. They were warring, they were visiting, they were taking their clothes to the Laundromat. Whatever they were doing, they were doing it in their canoes.

Most of us with any interest in the water tip our hats to tradition by owning a canoe. Here on the Tuckahoe, there probably are more canoes than all other kinds of boats put together. They are very seldom used because most of them are heavy to lug down to the water and all of them are work to paddle, but you see them everywhere, on racks or more often on the bare ground, turned upside down so that their aluminum oxidizes or their plastic distorts. It is assumed that any American can handle a canoe and get from one place to another in it, and there may be some truth in that, although handling a canoe well is surprisingly tricky. Still, when your cousin and family come for a visit and you tire of the kids whining and breaking things, you take them out and put them in the canoe, perhaps with a word or two of instruction. Usually nothing bad happens, and they get a lesson in their heritage.

We tend to think of Indian canoes being made of birch bark—white birch, in fact. But paper birch is choosy about where it grows in sizes big enough for canoe skins, and many of the Indian canoes were dugouts, while others were skinned with hickory, elm, or other bark. The European conquerors found these boats handy and at first copied the Indian building methods, but eventually came clinker and then canvas-over-carvel versions and finally metal and plastic ones. For one-offs, cedar or pine strip-planking with light fiberglass outside and in has been popular for many years now. Curt Read, Franny's husband, has been building a kayak by this method for about two years. Of course, a kayak is harder to build than a canoe, having sharper ends that are harder to glass and also a deck that is as much work as a hull to construct, to say nothing of joining the two together. Nonetheless, he has put as much money into this boat as would have bought him an Old Town and as much labor as would have bankrupted the Old Town Canoe Company a long time ago.

Joe and Dee Bonner, across the road from us, wanted a lightweight canoe. Joe had had a big heart attack and didn't fancy lifting even half an 80-pounder. They wanted to paddle it on the river here, and they especially wanted to take it in the summer to Maine, where they had been renting a campsite on a lake island for years. What kind of canoe could I build that they could easily carry?

A surprisingly light canoe can be made using Kevlar rather than fiberglass. A 14-footer, for example, can weigh as little as 40 pounds. However, all excess resin has to be vacuum bagged away, as is done in airplane construction, and because Kevlar is virtually impossible to sand, the boat is best made in a mold, so Kevlar's not a good one-off material. Kevlar is also as expensive as any boatbuilding material that I know, so a mass-produced Kevlar canoe costs about twice what its fiberglass sister does.

One technique that has been around for a long time and has some merit for lightweight canoes is canvas-over-stringers construction. Kits for such boats were once popular. The stringers were closely spaced to give a nearly round bottom, and the frames that gave them their shape were $^1/_4$ -inch plywood, but what made the boat waterproof was alkyd paint put on over the canvas. At first the weight wasn't significant, but when the paint aged, it cracked and leaked, so a new coat had to be applied every year, and eventually the boat had to be recanvassed or used as an anchor.

A modernization of this method and a very great improvement occurred to Platt Monfort. Monfort is a machinist by trade who worked himself up to being a missile engineer (without the engineer's degree); in his spare time he worked on nautical inventions. Monfort developed Git-Rot, "the original proven cure for dry rot," and is now heartily sorry that he sold the concept and that even his royalties on it have now expired, because apparently dry rot is still alive and well and Git-Rot is still selling.

Dry rot is the same organism as wet rot; it's just moisture-starved for the moment, but alive and looking forward to eating wood again as soon as it gets a reviving drink. It's hard to be sure that there isn't some moisture in dry rot, down under the surface; if there is, no petroleum-based chemical will mix with the moisture and penetrate to the front line, where solid wood is being attacked and consumed. Dave Carnell points out that common antifreeze kills rot spores, does mix with moisture, and thoroughly saturates rotting wood, whether wet or dry. Git-Rot is supposed to bond with the damaged wood fibers to make a new composite material that is stronger than the original wood, and of course antifreeze won't do that, so other steps must be taken after the antifreeze has dried. Perhaps that is why Git-Rot, which claims to do it all in one application, continues to sell so well.

Monfort next developed Fer-a-lite, a mixture of polyester resin, short-strand chopped fiberglass, and lightweight aggregate. Catalyzed, it comes to a peanut-butter consistency and was used to build whole kayaks, which weighed about twice what their wooden counterparts would have, and troweled on over wire mesh, to sheathe tired wooden hulls. It weighed just a little less than water so did not sink a

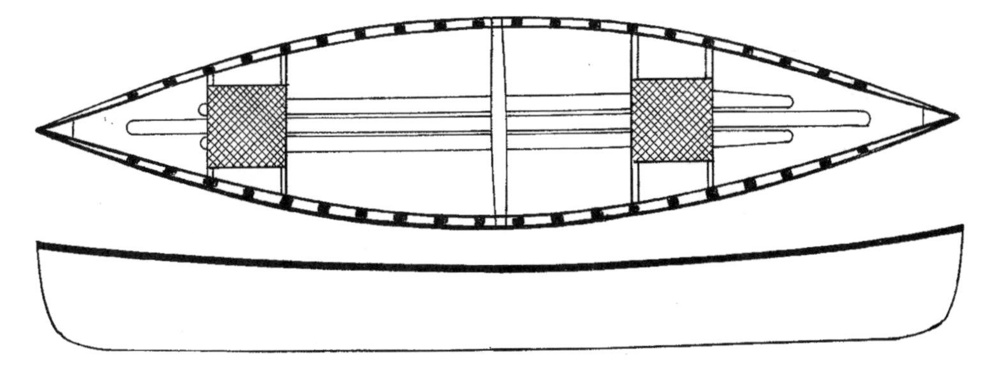


Geodesic Airolite canoe

The Bonners' Geodesic Airolite canoe in frame

hull too badly, and I can remember do-it-yourself boatyards where half the owners were in one stage or another of Fer-a-liting their hulls. More recently I have seen old wooden sailboats in the 40-foot range that have at last been discarded because the designs are obsolete: the overhangs too long, the cabins too small, the rigs too hard to handle. Old fiberglass boats have become so inexpensive now that fewer people are continuing with old wooden ones. But the Fer-a-lite sheathing of these hulls, applied perhaps twenty years ago, still looks like new and hasn't cracked.

Geodesic Airolite, which Monfort began developing in 1983, is surely his cleverest invention. One can speculate why he chose the name. *Geodesic* is an adjective meaning "made of light, straight structural elements largely in tension." It reminds us of Buckminster Fuller's geodesic domes, which were high-tech building designs for the low-tech builders of forty years ago. Airolite means nothing at all, according to all the dictionaries that I own, but it implies light weight, or light as air, which some of Monfort's designs very nearly are. The whole phrase trips pleasingly from the tongue, suggesting a very high-tech building method that we can readily master, as indeed we can.

There aren't many steps to the Geodesic Airolite building method. Molds are set up on a strongback, and stringers are let into notches in the molds and glued to plywood stem and stern. Ribs are bent to the inside of the stringers—and that is an enormous improvement over canvas-on-stringers because there the plywood frames were forever in the way, compartmentalizing the hull. Kevlar rovings are crisscrossed over the outside of the stringers on about 6-inch centers, and the whole structure is covered with Dacron cloth, which is then heat-shrunk. It is painted with polyurethane, which not only waterproofs it but also penetrates the fabric and sticks it to the Kevlar and wood.

Monfort has probably sold thousands of plans for these boats and nearly as many partial kits of the materials, which include everything needed to construct them except the wood and paint. He is successful for a number of reasons: first, his advertising is catchy. He is a prolific coiner of phrases, and he stresses the light weight of his boats and the simplicity of building them. Many of his designs are good and based on proven models from the past, such as Rushton canoes and Whitehall pulling boats. His plans are immaculately clear, with no detail left out and full-size patterns for every mold, even showing the notches where stringers go. An accompanying booklet tries to detail the building process, but writing is Monfort's one weakness. After I had read the booklet for the canoe several times, some steps still weren't clear, and I was glad when Curt Read lent me a Monfort-made video, which is homey but enlightening.

Monfort's scantlings are super light, and I might not have started on a Geodesic Airolite boat if I hadn't seen Curt and his friend Chuck build and use a couple of them. Stringers are $^3/_8$ -inch square. The keelson is $^3/_8$ inch by 2 inches, and the floorboards are only $^3/_8$ inch thick. The reason Monfort can specify such lightweight scantlings is that there are very many of these small elements, and they are glued together the way a model airplane is.

Monfort is a genuinely nice and helpful man. God knows how many hours a day he spends on the phone, answering his builders' questions, and when someone who lives not too far away is stuck, Monfort will even drive over and spend a few hours helping him out. I called him in Maine during the building of the Bonners' canoe because I had come to a problem myself. He said he didn't know the answer to my question but would think about it. The next night, after he had figured out a solution, he looked through his files until he found my order and letterhead, and he called me up, from Maine to New Jersey, to put me on the right track.

In addition, he knows how very reluctant most amateur boatbuilders are to spend money. He tells us we can build a box-section strongback using two one-by-sixes and two pieces of cardboard (at least 275-pound test) glued together. The molds can also be made of this cardboard, although he does say that some extravagant builders prefer underlayment. For the stringers he tells us that some people prefer not to use select stock and instead pick over housebuilders' two-by-fours, cutting the knots to waste and perhaps scarfing to make long enough pieces with straight grain. All the longitudinals in the canoe I built for the Bonners came out of 16 board feet of lumber. The price of select Douglas fir is about $2.50 a board foot. Is it necessary to save part of that money by pinching and scarfing and perhaps breaking some of the stock while bending it into place? A good many builders think so, and Monfort knows that.

For the first boat that Curt and Chuck built, they tried using the cardboard strongback and molds, just to see how they would work. They did not economize on the wood parts of the boat. As might be expected when using a strongback only 6 inches square, they had trouble keeping the molds properly aligned; and as the molds wanted to bend forward and aft, they had trouble with that too. They did succeed in building the first boat that way, but for the second they went to a more substantial strongback and plywood molds.

They also tried a glue that at the time came with Monfort's kits, although it had been eliminated by the time I bought mine. It was meant for attaching some small bits (breasthooks and the like), and it was the kind of glue that kids enjoy sniffing. Curt and Chuck could hardly help inhaling it, and soon they were giggling and the parts weren't fitting together properly, and it all seemed not to matter except that it was a great joke. Later they had to cut the parts out and glue in new ones with epoxy.

In 1994 when I was building for the Bonners, Monfort had two canoe designs, one 14 feet by 32 inches and the other 16 feet by 36 inches. As the 16-footer had a carrying capacity of 500 pounds, it seemed unnecessarily large, and however light it might be, it would be awkward on land and take up a lot of storage space. But the 14-footer with its narrower beam would require the paddlers to kneel, not sit, and in fact the study plan shows spreaders rather than thwarts in the boat. After talking the designs over with the Bonners, I bought 16-foot plans and shortened the boat 2 feet by putting the molds closer together. Monfort now sells plans for exactly such a boat.

Geodesic Airolite is not a good method for a professional builder but nearly ideal for an amateur. There are lots of hours in it because there are so many parts—too many for anyone to make a living cutting and assembling—but my, the work is pleasant! There is very little sanding and finishing, and as always, the steaming of the ribs is the most fun of all. Monfort says that if you don't want to steam, you may get away with using the greenest of sapwood ribs and banding them with steel on the outside (between rib and stringers) until they have taken a set. It sounds like more trouble than steaming and less fun, too. Any straight-grained wood from the temperate regions (not the tropics) can be steamed more or less successfully: walnut, oak, elm, ash, or my own favorite, sassafras. Sassafras is lighter weight than the other hardwoods but still holds fastenings well. It smells good. Unlike oak (and there's practically nothing I use oak for, if I can find another wood to do the job), its grain does not rise with steaming.

People tell you that you can't use long-cut, dried-out wood for steaming, but I've had no trouble with it. However, air-dried wood grows dry over a period of time, and steaming does not put moisture back into the wood but only keeps it from drying out while heating to flexibility. If hardwood several years old is what you have on hand, you should rip it to dimension several weeks before it's needed and let it soak in water, well weighted down, before putting it in the steam box. With air-dried hardwood that has lost its moisture only over time, soaking has always worked for me. Kiln-dried hardwood may have lost resins or other elements in the drying process and may not steam so successfully, but I have never tried using it.

Very many clamps are needed to hold the ribs against the stringers. If a clamp were used at every junction (and it's possible to skip a few), this 14-foot hull would require nearly three hundred. The clamps don't have to be powerful because the ribs are only $^{1}/_{4}$ inch by $^{5}/_{8}$ inch, but some need to be stronger than a spring-type clothespin. Mine were made from short, slotted sections of $2\,^{1}/_{2}$-inch PVC pipe, as shown in the photo. PVC cuts readily on the bandsaw although it doesn't do the

PVC clamps for glue-up of ribs and stringers

sawblade much good, so use an old one. I used the clamps again after the ribs were set and dry, when I marked the junctions of ribs and stringers, took the ribs out one at a time, epoxied the contact point, and put them back in. Obviously there are some three hundred joins just for these places, and it's beyond my patience to clean up all of them to the standard of furniture quality. But when you look inside a completed Geodesic Airolite boat, your eye is so busy with the thousand details and pieces that it does not linger on any single join or blob of glue.

The phase of building that scared me most was heat-shrinking the Dacron. Monfort sells the type of Dacron that covers airplane wings, where it must also be heat-shrunk, and it comes much wider than his boats, so it is put on diagonally, pulled taut, and attached to the gunnels. Then it can be shrunk with a heat gun, he says—but cautions we have to be clever with the gun because a little too much heat can melt the cloth, so better use a laundry-room iron. He tells us how to test the iron's heat on a scrap of fabric, and says that we must not iron wrinkles but iron beside them until they pull taut and disappear. I did try his heat-shrinking method, but ironing is not like any kind of boatbuilding I ever did, and it made me so nervous that in the end Carol took over; she made a fine job of it.

In addition to being 2 feet shorter than stock, the Bonners' boat has 2 1/2 inches less freeboard at bow and stern, which is what my study plan shows. Most rowing and paddling boats are too high in the ends. The freeboard is put there for looks, which it may give, but what it certainly gives is windage afloat and more awkward handling on land. A little sheer curve in an undecked boat is what we're all expecting, so I put 2 inches into 13-foot *Chingadera,* but Monfort shows 8 inches on his 14-foot SnowShoe canoe. The extra weight is insignificant because it's just a few more square feet of painted Dacron, but this extra freeboard in the ends adds nothing to seaworthiness. It's impossible to think of a situation where it would keep water out of the boat.

The Bonners were very excited about their canoe and came over often during construction to take pictures and talk. So did other neighbors, and what interested them most was the Kevlar roving, which looks like unraveled yellow string. They pronounced the word almost as if it were holy. Finally I said to a couple of them that the strands looked like a whore's hair. She said, "I don't know about that." He said, "We don't any of us know about that. It's just something we heard about."

The Bonners bought a pair of oak-framed cane seats for their canoe by mail. They are a couple of pounds heavier than the plywood seats that Monfort showed, but they do look pretty, and at 31 pounds the boat is light enough for them to manage. In fact, the light weight was a problem at first because light boats have very little stability until you're down in them, and at least once the Bonners cap-

The Bonners in their Geodesic Airolite canoe

sized at dockside. They got used to it, though, and did take the canoe to Maine several times, where it gave good service and brought many comments from other campers. Joe and their older son entered her in a Tuckahoe River canoe race once and won a prize. Carol and I tried her on a windy day and liked the way she tracked and handled.

This canoe is six years old now, and Monfort was once estimating that such boats might last ten years without reskinning, which would be a difficult job to do neatly because the old skin is stuck to the stringers and Kevlar by the paint. However, the Bonners' canoe is kept out of sunlight when not in use, and it really is painted (Monfort has an unfortunate weakness for varnish, even the best of which is a very inferior sunblock to the cheapest paint), so I think this one may last a good deal longer than Monfort's estimate. We shall see.

# Herreshoff Prams

Most places in South Jersey you can put down a mooring anywhere without buying a permit. The problem is then getting ashore, and that usually does require permission and payment. However, fifteen years ago a restauranteur near the mouth of our river decided that sailboats on moorings and the decorous comings and goings of their crews might induce his customers to linger for another drink. He opened his beach to tenders, and that is why I got into the pram-building business.

There isn't much merit in designing an 8-foot dinghy from scratch. It's been done so often that the good ones are well known, and they're all pretty much alike. What's called for is a boat that carries a heavier load than should sensibly be put on such a short waterline, that is dry in all but catastrophes, and that does not waste interior volume (for example, by having a pointed bow).

In the *Nereia* tender L. Francis Herreshoff got it about right. The extreme rocker is needed to carry even two people. The bow overhang and 30-degree bow-transom rake are the only ways to keep an 8-footer dry, unless a motor is clamped to the transom and the boat is made to plane, in which case a bottom with no rocker at all serves best. The 3-foot, 7-inch overall beam is narrower than usual, but it actually works out better for rowing. The worst design fault is the 10-degree rake of the stern transom, which robs the boat of a precious 2 inches of load-carrying bottom compared to a plumb transom.

The tender was designed to fit between the mizzen and cabin of Herreshoff's 36-foot *Nereia* ketch, where it must have made a fine fandango of getting in and out the companionway. My prams were meant to serve very much smaller boats—a 17-foot cat schooner, a Lightning, and several other daysailers—and were never to be taken aboard but simply rowed out to the mooring and left there while the mother ship was daysailing in the bay. They worked only fairly well, and the fault was partly mine and partly the materials I used.

At the time I was new to the area and eager to use local materials and building techniques that I had read about in Chapelle and elsewhere. Herreshoff's "dory" building method, which called for fore-and-aft bottom planking, certainly carried the New England taint, and I resolved that my prams would be cross-planked. That meant they could not have the outside seam battens with which Herreshoff hoped to keep his bottoms tight. His suggestion that if the boat was left out of the water, leaks could be avoided by using "laminated wood" (that's the closest he could ever come to mentioning plywood) I also dismissed as not local and traditional enough. Herreshoff's *Nereia* tender has sawn frames, two planks to the topsides, and no chine logs (undoubtedly the cedar-to-cedar join at the chine does leak in most boats built his way). I would depend instead on a single center frame, chine logs, three planks to a side, and sassafras frames steamed in, as was done in days of yore. I also put in a $^3/_4$-inch square batten down the center on the outside for plank fastenings, which culminated in a skeg that Herreshoff omitted, probably to avoid clutter when the boat was upturned on *Nereia*'s deck. I also rounded the transom tops, which Herreshoff had avoided so the tender would be stable upside down.

I built these prams two at a time, which was something of an economy, and sold them unpainted, which saved a good deal more. I hoped to build them more quickly as time went on and to make a stock model of them, perhaps even building them on spec and advertising them. But the second pair were less interesting to build than the first, and I doubt that there's a place for stock wooden boats in today's market.

Compared to an aluminum canoe or a fiberglass runabout, the labor saved in a stock-model wood boat is not very great. Masonite patterns could be made of the transoms and topsides planks and perhaps a few other parts, with bevels noted on them; but each boat would still need to be finished, and an aluminum or plastic boat needs no finishing. The greatest advantage of building in wood is that the owner can have exactly what he wants. The buyer of one of these prams wanted his to be 7 feet long, not 8 feet, with a mahogany top strake that he could varnish, and it was no trouble to accommodate him. Conceivably a stock wooden boat might catch the fancy of some buyer who doesn't know what he wants and won't know until he sees it. Usually such buyers want a plastic boat, with a jet engine.

Pacemaker and Egg Harbor, two local builders of big powerboats, stuck with wood a long time. Their hulls were built over molds with ribbands attached, and the shape had no tumblehome so that it could be lifted off the mold. The ribbing crew came first and steamed ribs over the mold. The carvel planks came down from upstairs, all cut to patterns and numbered so that there was no fitting. The planking crew screwed each one on, and both factories could turn out the hull of a 40-footer in a morning. A fiberglass 40-footer takes several days to lay up, because the resin gets too hot if put on at one go and the shape distorts. But even with this economy of speed, Pacemaker went bankrupt. Egg Harbor eventually converted to fiberglass but went bankrupt anyway and has been resurrected and gone bankrupt several times since.

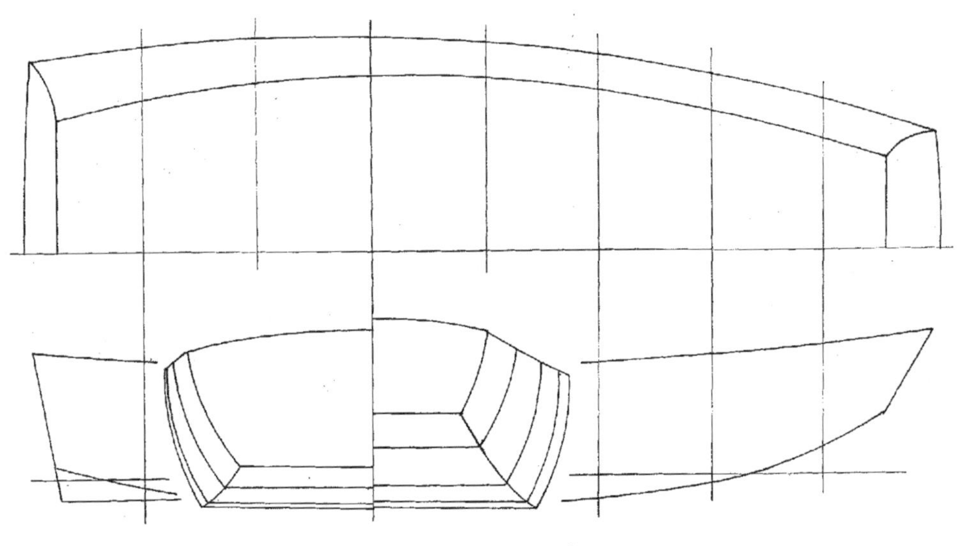

Herreshoff pram lines

One big improvement that I did make in Herreshoff's design was to put the oarsman's seat fore-and-aft. He had three thwartships seats in this tiny boat, with pairs of oarlocks at the forward and middle thwarts, a configuration that made it difficult for all but the smallest rower to keep the oar handles from hitting his knees on the recovery stroke. It often happens with a tender that a single oarsman goes ashore to pick up a passenger. With the fore-and-aft seat he can simply slide forward until his calves are about touching the center frame and move the oarlocks up to their forward sockets, and with the passenger on the stern thwart the boat's balance is adjusted to suit two people. My prams have no third seat in the bow, however; when carrying three people—a marginal enterprise in an 8-foot tender and only to be tried on short courses in flat water—the oarsman sits amidships, and the third person squats in the bow. I have been that third person and survived.

The four prams that I built went together without trouble, and they were presold. They had little touches of quality: bronze oarlocks, steamed walnut knees in the corners. Transoms, gunnels, and the sawn center frame were the decent Philippine mahogany that was still available in those days. For one I made a pair of leathered spoon-blade oars, which certainly was silly. A short boat like this should be rowed with short oars and short, rapid strokes to keep some way on her between strokes. Anyway, I don't think the pram with spoon oars ever was used. It was meant to service a 15-foot fiberglass daysailer that wasn't quite ready for the water yet. The daysailer was sitting on a couple of four-by-fours in the owner's yard. When he heard that a hurricane might be coming he put a garden hose in her

Herreshoff prams

and filled her to the gunnels. A neighbor told him that wasn't a good idea, but the boat sat that way for a couple of days. When the owner went to bail her out, he found that the bottom shape had conformed to the four-by-fours. Emptied, it didn't come back. This was by no means a new daysailer, and I hadn't known until then that fiberglass would do such a thing.

The other three prams more or less gave service. However, they always leaked. One owner thought his would look good with a black bottom, and that one leaked spectacularly because black sucks up the sun's heat; but the others leaked also, too much to be pleasant boats to use. The centerline batten on the outside was enough to keep the planks from working against each other and spewing caulk, but the boards shrank and swelled with the weather. They weren't the same virgin-forest cedar that Chapelle confidently assumes will be used. They were second growth, which means that the wood is less dense with annual rings farther apart and more tendency to shrink and swell. They were plain-sawn, not quarter-sawn, so the rings were at any angle to the flat of the board: for minimum shrinkage (and cupping) the best planking is quarter-sawn, with rings perpendicular to the flat.

So the prams would be launched and rowed out to the mother ships. A bit of water would dribble in from the bottom seams, which were thoroughly dried out from their stay on the beach. It wouldn't seem too bad, provided that sweaters and the paper bag of snacks were kept carefully out of the bilge. But after the owners came back from an afternoon of daysailing, the pram might have several inches of water in it and would have to be bailed before it could give service. That's not a boat that a builder can take much pride in, and I've since wished I'd taken the advice of Mr. Herreshoff and made the bottoms of "laminated wood." I sold the prams for less than marinas were then charging for one year's mooring, so most owners did get their money out of them. Then the restaurant changed hands, and the new owner decided that a fiberglass cruising sailboat drawn up on the beach and converted to a bar would be a cheaper and less troublesome entertainment for his customers, so the fleet was banished. I don't know what happened to the prams.

## *Dandy* Dinghy

Here is a dinghy that was worth designing from scratch! My last book, *Multihull Voyaging,* showed a study plan of her and told something about the design but nothing about the building. Here you have complete plans, and in addition, we have six more seasons of experience with her, so I am able to tell you a bit more about what she can do.

This dinghy came about because Carol and I were tired of inflatables: they are nasty to row, wet to sit in, and hard to patch if they're punctured or abraded near a seam. The *Dandy* dinghy was designed purely as a rowboat, and she is not a good model for a sail or a motor. She was intended to float on her marks with my 160 pounds rowing and Carol's 110 pounds in the stern and perhaps a few grocery bags or water bottles on each side of me. She is so carefully calculated that if we merely change places she drags some transom and Carol at the oars has some trouble keeping her on course. She is a very specialized boat, and that is why she is so good at what she does.

I wanted a flat or nearly flat bottom amidships for stability on 3 feet of waterline beam and veed sections forward and aft to produce decent rowing lines. Such a shape was tried on sharpies in the nineteenth century, but the design that had been intriguing me for years was a 37-foot scow schooner, taken off by Chapelle in Galveston in 1941 and admired by him so much that her plans are included in his *American Small Sailing Craft,* although she was by no means a small boat. It may seem odd that such a big boat should be the inspiration of such a small one, but the dinghy's lines are not just a scale-down of the schooner. She has more depth in relation to her length, virtually no rocker for several feet amidships, and of course a bow overhang with 30-degree bow transom, as pram dinghies must have to be dry.

You might think that with frames 2 feet apart and almost 18 inches between chine log and keelson, something heavier than the 5-millimeter okoume used in *Chingadera* would be needed for this bottom. It isn't, because the bottom panels are curved athwartships so that the plywood is a compound curve, which stiffens it enormously. Stiffness is not to be confused with strength, and a sharp stone or even a nail in a balk of timber could pierce the dinghy's bottom as readily as it could the kayak's. Stiffness means resistance to flexing, ability to hold its shape and not oil-can; as for sticks and stones—the oarsman had better keep his eyes open (and in the back of his head, a lamentable feature of rowing).

On *Dandy*, our new catamaran, we have only 6 feet by a little over 3 feet of deck space for dinghy stowage, so clearly this tender would have to come apart. The usual break is under the oarsman's seat amidships, but for weight, windage, and dryness, I wanted a decked forward section of low freeboard, so it seemed more logical to break the boat there. The bow section has 200 pounds of buoyancy, and if it ever submerged (it hasn't, so far), the coaming and lip of the aft section might still stop the aftershock. (However, in a dinghy or kayak we swing to take wakes on the beam, where they bobble us annoyingly but harmlessly.)

The boat cannot be taken apart and put together in the water; such tricks require very sophisticated hardware. The bolts and wing nuts shown make assembly quick, but it's best to carry a couple of extra wing nuts in the tool box of the mother ship. The bow section fits comfortably in the stern section and is tied down with a rope before the boat is turned over. The plastic oarlocks are hardly traditional, and they can flex enough to pop the oars out, especially when backing up. But they are lightweight and wonderfully quiet, and they show no sign of wear after eight seasons. The steam-bent knees shown in the plans could, of course, be sawn knees, but why deprive yourself of the pleasure of steaming?

We did not install the skeg shown, and she would track straighter if she had it, especially with only one person aboard. Like Herreshoff with his *Nereia* tender, we didn't want a skeg in the way when the dinghy was upside down and we were handling sails. We do have a $^3/_4$-inch-square keel from bow to stern that fits into a trailer roller on one bow of the catamaran. Of course, the veed bow and stern give her considerable tracking ability compared to a flat-bottomed boat. Perhaps when she wanders she is being pushed a little faster than an 8-foot waterline ought to be.

The oars should stow within the boat (oars can easily be lost at a dinghy dock when other people, while trying to find or make a slot for their tender, drag their painter across your boat, catch the oars, and inadvertently float them away). The very wide blades are meant to compensate for shorter-than-ideal oar length. To secure the boat against theft we only take out the oarlocks. People are not likely to steal a rowboat anyway, with so many outboards to chose from. If ever we come back to a dock or beach and find the dinghy gone, it will probably be visible in the distance with a couple of kids paddling around in it.

As for looks, it is often said that beauty is in the eye of the beholder. I design boats to do their job and adjust the looks afterward. The sheer height of the dinghy's aft section is dictated by the need to have the oarsman's seat a little higher than his feet and the oarlocks 8 inches higher than that. It is straight and parallel to the waterline. The false sheer, a continuation of the guard rail of the bow sheer, does stiffen the topsides, which are not a compound curve. In our boat its importance is heightened by using two different colors: bright yellow below and white above. Perhaps greater contrast would be better yet, but she is painted from the same cans as *Dandy*. I fancy she looks a little Venetian, or anyway Italian of some sort, and she gets compliments wherever she goes, so there must be other beholders like us.

To build the *Dandy* dinghy, you need two sheets of 5-millimeter okoume and an extra piece of something to make the forward deck. This could be 3-millimeter okoume, which is light and cheap, for there's no strain on the deck; foolishly I used

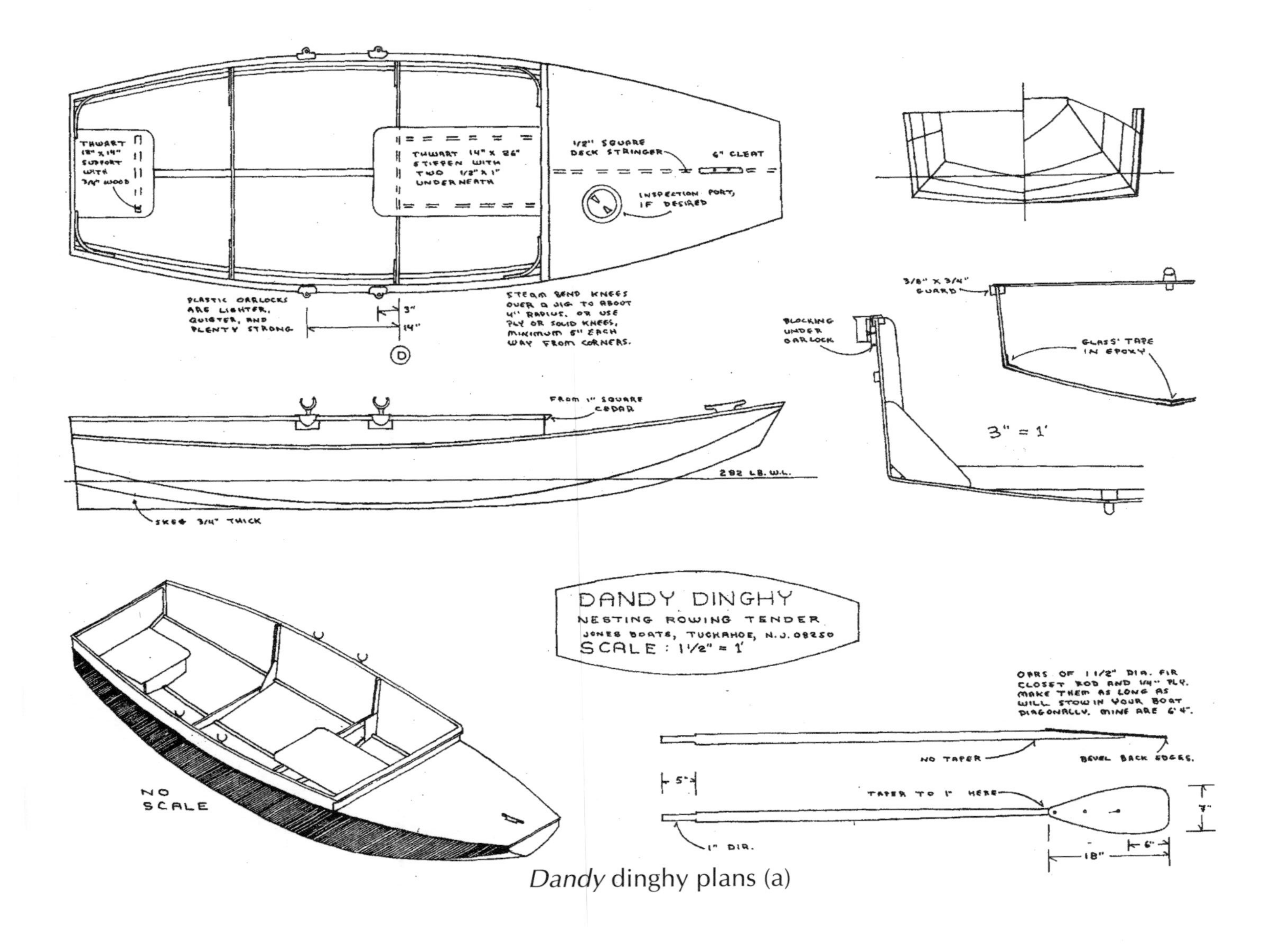


*Dandy* dinghy plans (a)

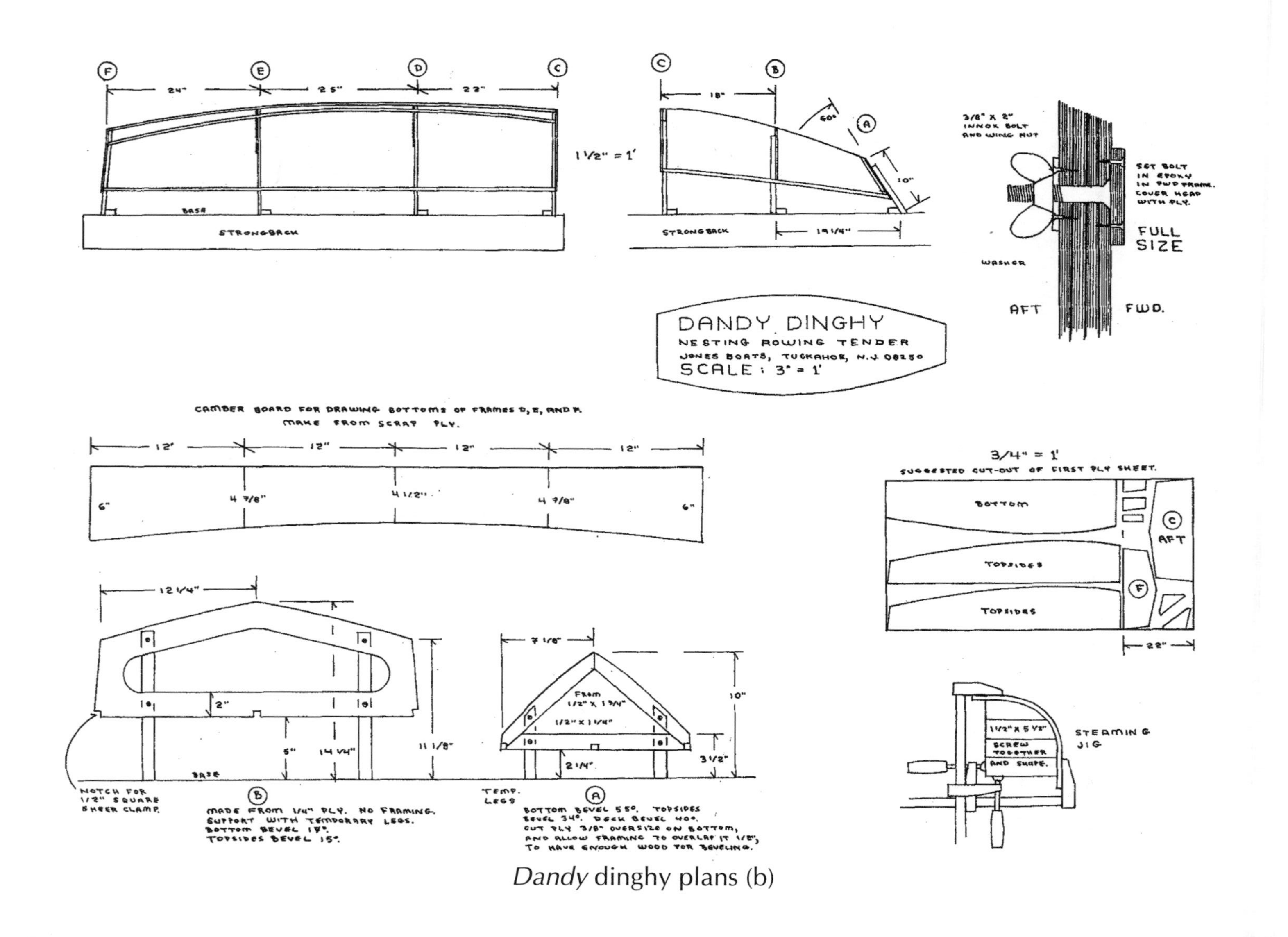


*Dandy* dinghy plans (b)

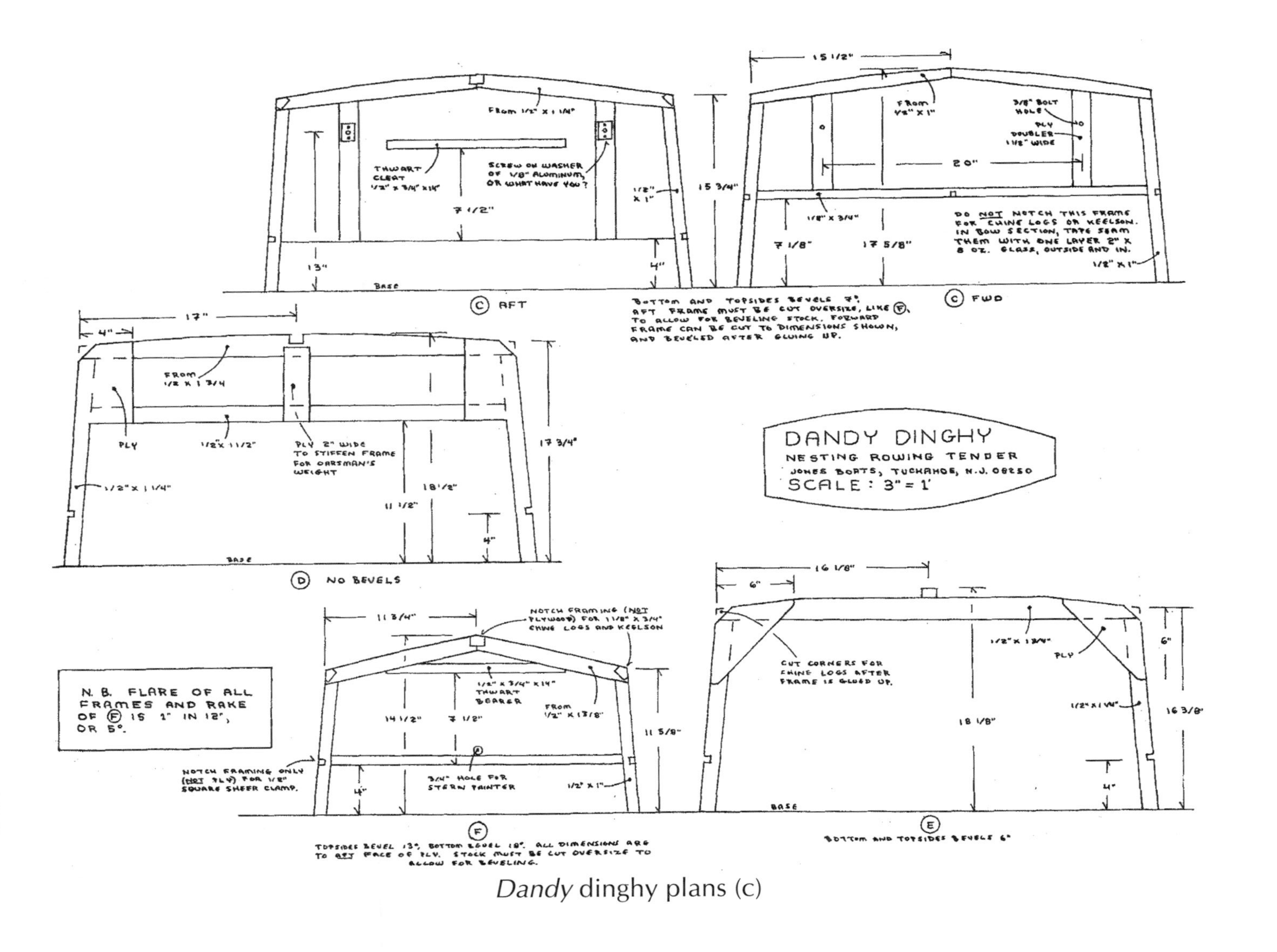


*Dandy* dinghy plans (c)

Carol rowing the *Dandy* dinghy

$^1/_4$-inch marine fir because a piece of about the right size happened to be lying handy, and it adds 3 pounds to our boat, which could have weighed just 50 pounds. If you are entirely confident of your joins you can delete the bow inspection port. I do look in mine once every couple of years but have never found anything except a little sawdust.

You need 12 board feet of Douglas fir or mahogany. To keep weight down, each stick is no bigger than it need be, and in most plywood boats the longitudinals where two pieces of plywood meet, such as keelson and chine logs, are $1\,^1/_2$ inches wide. My dinghy uses $1\,^1/_8$-inch stock, called $^5/_4$ stock in the lumber business, and these narrower longitudinals have held up fine for eight years now, despite the beating that a dinghy inevitably gets around docks and being hoisted aboard and broken down.

I did use lauan, but even eight years ago the quality was poor, and it would be abominable to use now unless you found an old piece or two in the back of a lumberyard. Lauan, or Philippine mahogany, suffered a wave of popularity in the years after the Second World War. It was used for house trim as well as boats, and you could buy it in various milled shapes: quarter-round, ranch, ogee, and so on. It was clear and less expensive than clear white pine, and some of it was so red it could almost be mistaken for Honduras mahogany, although the pores were bigger. However, nine different species of tropical cedar were classified as lauan, and some were nearly as white as pine.

The Filipinos cut it industriously and cut anything in its way as well, and last time I heard the islanders had felled 95 percent of their forests. The wood we are offered today is still called lauan, but the name actually refers to any of various kinds of tropical hardwood from Indonesia. Most of it has pinworm holes, and the wood tends to break on those holes if it's bent. It really doesn't want to bend at all. It is heavy and not dimensionally stable, and who knows what decay resistance it has, because no one tells us the actual names of the species. It is not suitable for boatbuilding.

Honduras mahogany is still available and still good, although it does cost a dollar more a board foot than what is now called lauan. I don't like what they are doing to the tropical rain forests where they cut it, but I don't like the growth of the world population either—or a good many things that go beyond the scope of this book. Douglas fir from our own North American rain forests perhaps is cut in a more responsible manner. Select grade is good for boatbuilding, although it tends to check and run sap and it doesn't hold paint well. Much of it is sapwood, and there's little change in color between sap- and heartwood; but I have found that fir sapwood lasts better than sapwood of some species supposed to have greater decay resistance, such as cedar. Increasingly I'm relying on fir.

For gluing, all brands of epoxy are good, although the working qualities vary somewhat and the prices vary phenomenally. Because plywood end grain absorbs moisture like a sponge, it should be coated with unthickened epoxy. I don't recommend using epoxy as a primer paint or for "saturation," as the advertisements say. Any good it does is by limiting the flow of oxygen to the wood. Dave Carnell's experiments have shown conclusively that epoxy only retards the way wood absorbs moisture, no matter how many coats are applied. Epoxy is more likely to keep moisture in. Epoxy resin is not self-leveling as paint is, so it must be sanded after it dries both to level it and to get rid of the amine blush, a chemical that rises to the surface during the cure and keeps the next coat you apply from sticking. Paint, whether latex, alkyd, or polyurethane, is designed to coat wood, to shield it from moisture and sunlight, and to last. I would as soon paint my boat with ketchup as with epoxy. Paint contains two elements: stickum and color. Primer is mostly stickum, and top coat is mostly color.

The two sections of the *Dandy* dinghy are built on a strongback, which needs to be only 6 feet long. The bottom futtocks of frames D, E, and F are traced with the camber board, which can be made of any scrap that's lying around. In the bottoms of frames A, B, and C the camber is greater: $^1/_4$ inch in each futtock, regardless of length. Throughout the bottom the camber works to stiffen the ply, but forward it also works to make the futtocks conform to the thwartships curve that the plywood wants to take when twisted. That is called conical development.

The idea is that any sheet goods—steel or plywood or cardboard—can be bent as if around the surface of a cylinder, in which case there would be no twist and so no sectional curves. Or it can be bent as if around the surface of a cone. To predict the sectional curves arising from that shape, you need to make two drawings (plan and profile) and locate the apex, or the pointed end of the cone. The two drawings must be reconciled, which often requires adjusting the line of chines or keel.

One of my friends is currently planking the bottom of a 60-foot aluminum sea sled. He drew the lines himself because he didn't want to pay the Hickman estate the royalty due on this patented concept. He drew the sections straight, and the keel profile—the line that most often needs adjusting—he drew to suit his fancy. After winching up an 8-foot-by-20-foot aluminum plate and seeing that it could not be forced against the frames, he decided to rip the plating and make a kind of strip-planking out of it.

With $^{1}/_{4}$ -inch or even $^{3}/_{8}$ -inch plywood you can take liberties with conical development. The thinner the ply, the greater the liberty that can be taken, and that is exactly what tortured plywood means: the panel is tortured into a shape that is not the surface of a cone or a cylinder. A designer who is also a builder develops a feel for what he can get away with and what he can't. Whether tortured or twisted, sheet material with a compound curve is stiffer.

In the dinghy's bow section, the curve of chines and keelson are too great to allow wooden longitudinals, unless steamed or laminated. Instead, I've drawn them for tape-seaming, like *Chingadera.* Frame B is also attached to the planking with glass tape. Unless you're picky, the inside taping needn't be sanded at all.

Putting the false sheer on the main hull should be the last job before painting, when it is possible to bolt the two sections of the boat together and judge the curve by eye. The sheer is drawn on the plans, and you can take offsets from that and clamp the $^{3}/_{8}$ -inch-by-$^{3}/_{4}$ -inch strip about where it will go. Then you have to stand back and eyeball it from both bow and stern and make the small adjustments in the curve to produce a fair and pleasing line. This eyeballing of curves is as much fun as steam bending, so don't rush through it. You may even want to bring in a friend and talk it over, because it often helps to have one person scrutinizing the line while the other adjusts the clamps. This entertainment must go back at least as far as the Phoenicians.

The *Dandy* dinghy has given us wonderful service, and it excites more praise from strangers than does *Dandy* herself. Stowed just aft of *Dandy*'s forward beam, it is inconspicuous, and many people assume it is part of the mother ship's structure. In the water it catches all eyes. Catamarans can seem threatening to people who haven't tried them, but everyone can imagine rowing this little boat. Boarding her, we are stepping down more than 2 feet, but even the first person can step anywhere in the bottom and feel perfectly secure. Stepping on the seats is not recommended unless there is another person already in the boat.

We have rowed three people in her, which clearly overloads her. The oarsman sits amidships and the third person sits on the splash guard at frame C, facing aft. One evening in Newport, Rhode Island, we thought we had an impossible mission for her. A guy who lived nearby had recently bought plans for my trailerable catamaran Brine Shrimp (described in chapter 5), and he wanted to bring his wife over and talk about it. They were to come down to the beach at the south end of the harbor. Looking out and seeing them dimly in the dusk, I exclaimed to Carol, "Christ, they've brought a big black dog with them!" But the dog turned out to be a big black knapsack loaded with sweaters for them and dessert for everyone, so the dinghy was saved.

# Drift Boat

Some years ago George and Janice Betts bought a cabin on a Maine lake. It's remote; to reach it, you have to leave the car and hike in quite a distance. Before long they met a neighbor who had built himself a drift boat, and George was invited to try it. He liked it so much that he gave the owner a deposit to build a second one for him, and the owner generously said that until the second boat was built, George could borrow the first whenever he liked. A year or two later the owner had spent the money and the second boat wasn't started, so George paid him another deposit, which had the same result. Meanwhile the first boat was getting pretty beat up and leaky, so the Bettses turned to me.

I've heard of a similar case with another Maine craftsman, that one about a copper weathervane in the shape of a mermaid. The mermaid buyer was philosophical in the end, just as George was. Work is scarce in Maine, and although the cottage industry of taking deposits isn't hard work, it must require planning and some psychological insight, because some of us would not be such understanding customers as George and the mermaid buyer. I don't know what the next step would be, if the buyer chose to make a row about it, but presumably that has been foreseen and planned also.

Drift boats evolved in the Pacific Northwest, but their popularity is spreading east. Neighbors recently home from a vacation in Montana, Wyoming, and the Dakotas say that you "see those boats in backyards or trailers just everywhere out there." They haven't reached the mid-Atlantic states yet, because we would be seeing them on trailers. These boats *must* be trailered because they are designed to go down rivers, not up them, and to survive those runs the boats are far too heavy and cumbersome to be transported in other ways or to be rowed in flat water.

Drift boats are used for fishing and perhaps for short general excursions, but they are meant for white water. I have never tried it, although I do have considerable experience with white-water canoeing. A drift boat with its immense rocker can turn faster than a canoe. Its meaty construction survives harder knocks. Its wide bottom and flared topsides make it less likely to capsize, but the same great width means that it needs a wider space between rocks to get through unscathed, and it cannot be moved sideways in the water, as a properly handled canoe can.

Nevertheless, these boats need a skillful oarsman, and they are nothing akin to the inflatables in which "guides" take a half-dozen tyros down white-water rivers, paddling occasionally but in general expecting to bounce off the rocks. In the Poconos there are even special days when the dams are opened to make enough water for these "expeditions," and one wonders how much they differ from amusement-park rides.

George has explained to me how he handles his boat, and when he runs rapids he is using judgment, coordination, and strength, just as one would in any other serious sport. He sits facing forward and pulls on the oars in the usual way, so the boat is moving aft through the water, although the rowing can't keep up with the current and the boat is moving forward over the ground. The idea is to slow the boat down and give the oarsman time to make decisions because a white-water river is never the same twice. A difference in water flow makes a difference in current strength and also in which rocks can be floated over and which must be avoided. In fact, it was at exceptionally low water that George finally holed the boat that I built him. He came upon a pointed rock he had never seen before—from the way he shaped his hands to describe it, it was more a spear than a rock and just below the surface—and he hit it sideways and amidships. It came right through the $^3/_8$-inch seven-ply sapele plywood.

In addition to rowing backward to slow the boat down, George uses a stone anchor of about 25 pounds hung from a crane about a foot off the transom. He lets it go with his foot. He asked me to build this apparatus for him, but I didn't really understand it and he was in a hurry to take the boat up to Maine. I have seen it only in photographs and still don't know exactly how it works.

The boat was built from plans provided by a designer who has been in business many years and designs every kind of boat that you can imagine. The plans were disappointing for what they bothered to show and what they didn't show. There was, for instance, a full-size frame pattern, which hardly seems necessary for a plywood dory, but no indication where it went. The design was made thirty or more years ago, when good Douglas fir marine plywood was inexpensive and commonly available in any length you might want. Fourteen-foot and 16-foot sheets were called for with the greatest abandon and drawn with no thought to waste, so that several sheets left me with scraps 8 feet long and 22 or 23 inches wide that had no further use in this boat. Nowadays when you pay a premium of 25 to 50 percent for oversize sheets, you are likely to scarf or butt-block instead. And you are likely to look for designs that cut the ply to less waste. But perhaps for the banging around that drift boats take, marine fir is still the best plywood.

For years I used khaya plywood for everything, but it is not imported at present, so the choices are fir, okoume, and sapele. A 4-foot-by-8-foot sheet of $^3/_8$-inch marine fir weighs 32 pounds and costs $46; okoume is 25 pounds and costs $94; and sapele, 38 pounds and $135. George's boat is sapele, with a $^1/_2$-inch bottom and $^3/_8$-inch topsides, finished to my usual standard, which is better than house quality but not as fine as furniture. It looked nice. The first wake-up came when George dropped the boat off the trailer before he even got it to Maine and thought it was a big joke. And then there was the rock.

Even when it was good, fir ply was likely to check in sunlight. It always had some of those football-shaped patches where knots had been cut out, and the patches

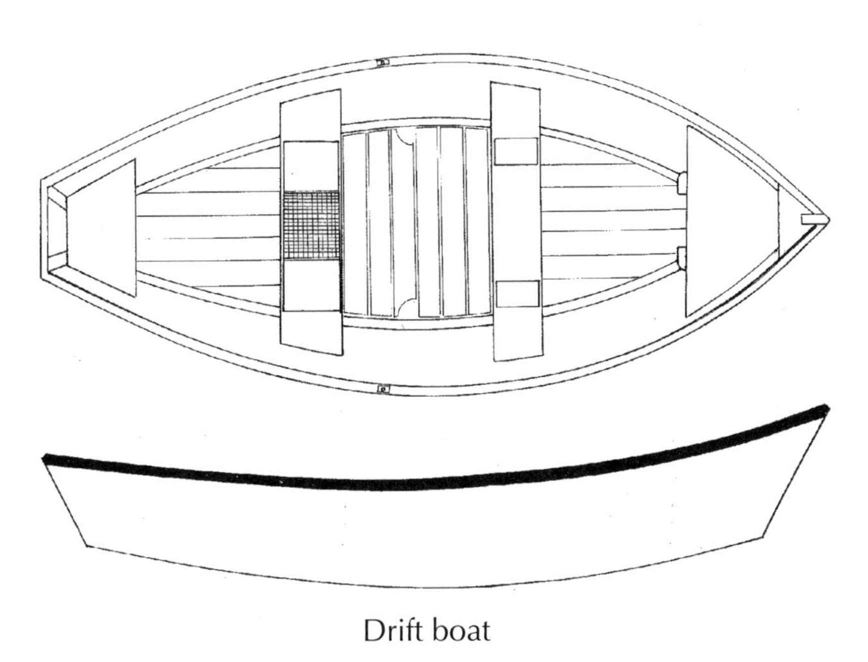


Drift boat

always popped out sooner or later. It had some "synthetic filler" (the American Plywood Association's euphemism for putty) where veneers had cracked in being flattened. Lately it has very much more of all that and is likely to have some voids, overlapping veneers, and so on. But when new, fir plywood is nearly as stiff and strong as sapele, and perhaps for this one type of short-lived boat it's still the best choice. George says that the drift boat he borrowed for several years, which is now six or seven years old, is about done. It's fir, but would it be in better shape if it were sapele?

Although stiffer and harder than fir, sapele plywood is not especially pleasant to work with. Like fir, it has a tendency to splinter along cuts, and the grain picks up if it's planed in the wrong direction. It does have good rot resistance, and the sheets are excellently made. Okoume ply is easy to work. It has less rot resistance, but that hardly matters in a drift boat. It is considerably less hard, stiff, and strong than sapele or fir, which matters very much for this application. Some drift boats have shock absorbers of dense plastic $^{1}/_{4}$-inch thick screwed to the outside, and the plans for George's called for meaty chine logs outside and in, but George didn't want those additions.

Building the boat from such peculiar plans would have been difficult for a first-time builder, but experience helped me to extemporize. I used two molds, a stem of $^{3}/_{4}$-inch ply with lumber packing pieces on each side, and a beefy transom.

I set them up on the strongback, ran the longitudinals, and traced the planking from them. With no twist anywhere, it wasn't tricky work. The plans called for glued-in floors of boards 5 inches by $^3/_4$ inches and covering half the bottom. On top of them George wanted thwartships cedar floorboards to keep water off his

George Betts in his drift boat on the Kennebec River, Maine
(Photo courtesy Janice Betts)

feet, notched out to give him footholds and ready to take the hardware for the stone anchor.

In fact, George wanted a very different layout from what the plans showed. Owners often do, so it's best to mock up the accommodation—the seats, the decks—using any kind of scrap such as cinder blocks and old two-by-eights and go no further until the owner has tried and approved it. George wanted the foredeck down about 8 inches from the sheer and a pair of little knees on its aft edge, against which he could brace his own knees when fly fishing. He wanted the two thwarts 3 feet apart, not 5 feet, and the thwarts themselves were to be relieved with limbered, open-topped boxes to hold small bits of fishing gear. The oarsman's seat was not to be plywood but a grid of ropes that would give him a little springiness and would shed water. No doubt he wanted this arrangement because he was used to it in the borrowed drift boat, and conceivably the arrangement shown on the plans would be better. But it *is* George's boat.

He wanted the whole boat varnished, inside and out. We don't incline to do that with our own boats, but varnish does touch up easier than any other finish, and it hides hammer dings and other little scars that show up wonderfully through paint, especially if the paint is glossy. It saves time and effort to paint a boat any one color and not have to cut or mask. Some people believe that the whole point of a wooden boat is to see the wood. A nephew of mine keeps a big fiberglass sailboat on Lake Erie and tied to it an inflatable dinghy that is not bottom painted and is not taken out of the water from spring until fall. He told me that for years he had wanted a wooden dinghy, varnished, but had concluded that it would be too much maintenance. I asked him why he didn't then have a painted wooden one, and he looked at me as if I were simple.

We tried the drift boat out on the Tuckahoe River, and George was pleased. I found it clunky, and certainly no one should have such a boat unless he plans to run rapids. This one must weigh 350 pounds. The waterline is short, and the beam and rocker make as much turbulence as any dory possibly could. It needs very long, cumbersome oars because of its beam and freeboard. George has rowed it some epic distances upwind on Maine lakes, but he has more strength and stamina than most of us, and he is single-minded.

Luckily his wife, Janice, is a very good photographer, so before the boat was started I had a chance to see how it would be used in her pictures of the borrowed one. Since then there have been very many pictures of this one in white-water action and also fishing in quieter stretches, rowing the dog around, and on an excursion with the grandchildren. It's not my kind of boat, but clearly it's his, and I'm glad he likes it.

The one incident not photographed was the encounter with the spear-shaped rock because Janice was in the boat at the time and was thrown into the bottom, suffering a mild concussion. George's immediate concern was for the boat, she says. He has patched it, but he says it still leaks a bit, so perhaps it will be back here for proper repair someday. Certainly there's enough scrap sapele ply in the shop to do the job.

2

# Daysailers

A cruiser carries you farther, a runabout gets you there faster, and a muscle-powered boat is more reliable. But a daysailer gives more pleasure than any other kind of boat for the time you have allotted to spend on the water. However, it must be a good daysailer, and the skipper must know how to sail it.

Opinions vary widely as to how sailing is accomplished. For eleven years Carol and I have been racing El Toro prams, and when she returns to schoolteaching after weekend regattas she often is asked how the races went. If the regatta has been canceled by a gale, some of her fellow teachers are highly amused. "Too much wind! But it's supposed to be a *sail*boat. How can there be too much wind?" Bear in mind that all these people have college degrees from somewhere. The way a cruising sailboat is managed is equally risible to them, because it seems self-evident that the stronger the wind, the more sail you put up.

Most people who want a daysailer have a better understanding than that, and many have even gotten beyond the notion that you go "where the wind takes you." But few fully envision the concentration and hard work sailing requires and the long, slow learning process. It is often thought that mastering sailing is about as difficult as learning to drive a car, or perhaps even easier because there aren't curbs to stay within; and it is true that a semi-skilled sailor can get into a boat and make progress in certain directions, especially if the wind doesn't shift. But I have been with many and many a sailor—some of them former sailing instructors at summer camps—and witnessed them tacking back and forth, back and forth across the Tuckahoe River and making no progress against the 1-knot tidal current.

Learning to sail takes about as long as learning to fly an airplane if the instructor is good. Sailing mistakes are, of course, less heavily penalized, but in the end it amounts to the same thing: you don't get where you planned to go. America issues no pilot license for sailing, but the U. S. Sailing Association now does certify instructors, and you should think twice before taking lessons from a teacher less qualified. After the lessons are over skills still need honing, and the best way to do it is in one-design racing. There you see boats identical to yours going past you. What are they doing that you aren't? Sometimes you can figure it out, and if you're sailing with nice people (as is always the case with El Toros), they tell you. Some may say that sailing is for relaxation, not for speed, but the wind is feeble propulsion for a boat compared to oars or a motor, and even the best sailboat does not go in one-quarter of the directions that you might choose. If hobbled further by the skipper's ineptitude, the voyage becomes so laborious that it won't be repeated.

The boat must be as handy as the skipper is. The rudder must turn her through the wind before way is lost, unlike a great many catamaran daysailers, which are more like missiles than sailboats, elaborately pointed by backing and filling, and then fired. The boat must make no more leeway in a decent breeze than the 5 degrees that are necessary to give the lifting action to the board or keel. The sail must set and draw well on a variety of headings and in a variety of wind strengths.

Here we come to a favorite economy of boatbuilders: Even if it is bought from Hong Kong, a sailcloth sail does cost money, and some builders may simply not have it, although more often they just begrudge spending it. Can a sail be made of other material? Certainly, and it's the very best place to save money, rather than on hull material, because a cheap sail can be replaced. However, there is now a fad for making a cheap sail out of polyethylene tarpaulin, and a much better setting sail can be made out of Tyvek. This material does have "Tyvek" written very large on every square yard, so the economy of having used it is no secret, but sailing friends will likely spot a polytarp sail almost as quickly.

Carol and I save money in various ways, but not on sails. We are on our third pair of El Toro sails, although we hardly use the boats except at regattas. The Tuckahoe Ten, which is the boat I sail most often, is ten years old and on her fourth sail. An Optimist sail proved too small. A larger one ordered from Hong Kong was too heavy a cloth to take a good shape in light air. The third used highly resinated light cloth from our El Toro sailmaker, and it wrinkled after several years of brailing. The present sail of light but less resinated cloth seems good, but if it dissatisfies me in some way I will not hesitate to order a fifth sail for the boat. The total cost so far is less than $1,000, or less than the cost of the air conditioner in our present automobile. How thrifty need one be about sails?

We are talking about the efficiency of sails here, so it may seem odd that all four daysailers in this chapter have sprit rigs. They do not point as close to the wind as Bermudan rigs, and as a boat only has a certain amount of stability, I doubt that enough extra area can be added to a spritsail to make up for that unweatherliness. But the important thing about a daysailer is not that it wins races over boats of different configurations, but that it feels lively and interesting in use, so as long as it gets you across the lake or up the river to windward and you are still interested when you get there. Again, oars or motor propel you faster, but you are sailing to have a good time—to pit your wits against the elements and get to some entirely unnecessary objective and home again. A sprit rig will do it.

I favor the rig whenever a boat is not kept in the water and the whole rig has to be taken down and perhaps apart between uses. Three of the four rigs in this chapter have brail lines, which when pulled bring the boom, sprit, and sail up to the mast and dump the wind out of the rig, which can be especially handy when the dock or beach is downwind. The spars are short, and if the sail area is less than 100 square feet, the rig can be assembled lying on the grass and stuck into the boat in one piece. A spritsail needs more adjusting than a Bermudan sail, both for the day's conditions and while under way. If the stanliff, which holds up the heel of the sprit, is too loose, the sail creases from throat to clew. If it's too taut, the sail creases from peak to tack. Really, a different stanliff setting is needed for upwind and downwind work; for this reason, the stanliff in John Guidera's Melonseed, described at the end of this chapter, is led down to a block on deck and aft to be near the skipper's hand.

To some extent a stanliff also needs adjusting on port and starboard tacks because the sprit comes down one side of the mast and so has more compression force on it on one tack than the other. That is the greatest shortcoming of the rig. The traditional solution is to make the mast rotate, but I have never had any luck with that, despite very careful shaping of all components, low-friction materials such as a metal mast step, and frequent silicon spraying.

If boomed, a spritsail can profit from a vang, although it doesn't need one as much as a Bermudan or gaff sail. A boom definitely is better downwind in strong conditions, when all single-hulled sailboats take to rhythmic rolling; an unboomed sail pumps and aggravates the rolling. However, the Melonseed doesn't have one because John didn't want one, and the Dobler, one of the other boats described in this chapter, doesn't either because its owners were new to sailing and I thought they'd have enough aggravation without cracking their heads on a boom.

In a daysailer under 16 feet, one sail is enough, and wire rigging is too much. For a while the English thought that a trainer for kids should have all the complications of a larger boat, and the 11-foot Mirror dinghy had a spinnaker as well as a jib. But now in England as elsewhere the single-sailed Optimist pram is becoming the preferred trainer, and kids are taught to concentrate on getting that one sail to set right. At regattas where different classes compete, we often see Albacores and Lasers sailing together. The Albacore has every conceivable rig complication, and the Laser has only a single sail on her unstayed mast. They are going exactly the same speed always, and both the hulls are from excellent designers, so apart from keeping two people busy, what is the point of the rig complication?

The simple rig is used more often because it is easier to set up and take down. It gives fewer headaches because less can go wrong, and more pleasure because it more often sails at maximum efficiency. This chapter shows four boats with simple rigs, although their hulls vary in complexity. They are the work of four different designers, and we'll start with my own mini-garvey.

# Mini-garvey

A sailing garvey is a kind of scow and a close cousin to the more glamorous round-bilged inland scows that were built in great quantities and very large sizes and were often seen skimming across midwestern lakes fifty years ago. The garvey shares the virtues of the inland scows: a light stiff boat that sheds wetted surface as it heels and maintains a neutral helm even when pressed well over.

Thwartships balance matters at least as much as fore-and-aft balance in maintaining a neutral helm, as we found out when we tried a biplane rig on *Dandy.* A scow's hard bilge means that the center of resistance moves to leeward along with the center of effort. Fore-and-aft, a scow's square shape remains symmetrical when heeled. One would think that a double-ended shape like a dory would also benefit from heeled symmetry, but the lack of form stability (the lack of buoyancy outboard of the centerline) guarantees such boats a ferocious weather helm.

Ancient boats—Viking ships and the "Jesus boat" recently pulled out of the muck in the Sea of Galilee—were usually double-ended because people simplistically assumed that you ought to end a boat the way you began it. Helm balance wasn't at issue because the rigs precluded bringing the wind forward of the beam. People also thought that a hull should be wider forward than aft, because that's how fish are. Well-shaped daggerboards and keels do have fishlike sections, but like fish, they are operating wholly underwater. For a hull that works on the water's surface and therefore makes waves, other shapes work better.

A double-ended shape is best for rowing or paddling. Long ago I built a transom-sterned canoe, calculating the weights and thinking myself very clever, and with two aboard she rested exactly on her marks. But as soon as she moved she made a wave train that put the transom below waterline and created so much drag that a leaky, beat-up double-ended canoe, shorter and heavier and discarded from a livery fleet, walked away from her with the greatest of ease. Whitehalls appear to be transom-sterned, but the transoms are too far above waterline ever to be immersed, and probably those boats were originally built that way to make more space inside for carrying and displaying wares. Originally they were bumboats, peddling their way through the fleet in New York Harbor.

Some people think that a double-ended sailing hull is a good looker, but that's its only virtue. Nearly all modern designers agree, and what is marketed as a double-ended cruising boat is almost always that shape on deck only and quickly becomes transom-sterned below the sheer. In fact, you might argue that only the term survives, because the pointed stern has become a round stern on most so-called double-enders, so all it does is add a foot or two to overall length and a dollar or two to dockage and yard bills.

A garvey really is a double-ender, although both ends are square. The style is still much favored in New Jersey where it originated but is now almost exclusively a powerboat shape. The crucial spoon bow, which is all that differentiates the garvey from that other scow the pram, suffers much at the hands of amateur builders unwilling to steam or laminate the framing members. I was not long in New Jersey—

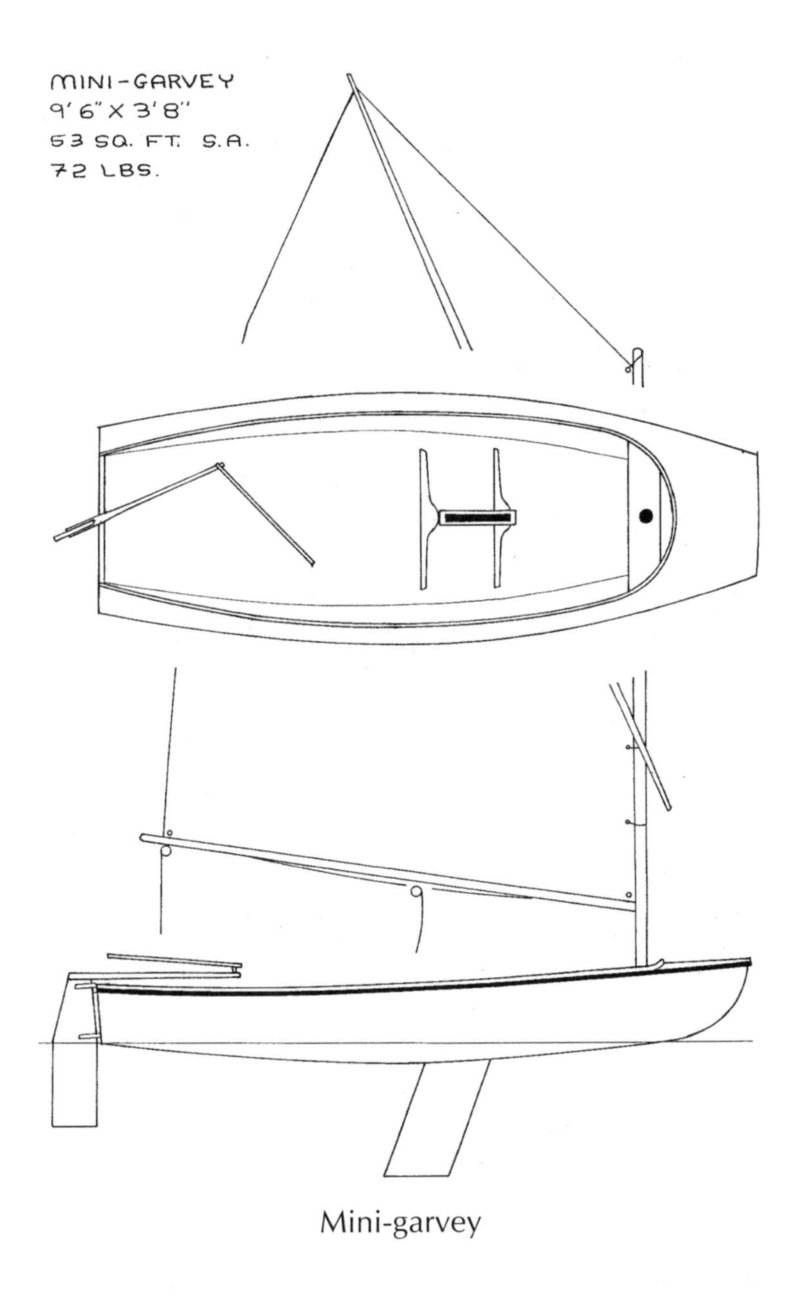


Mini-garvey

or "on the Jersey side," as they say in Philadelphia—before I had to have a sailing garvey, and of course she had to be planked with local cedar. The bow framing was laminated. Twelve feet long and with 90 square feet of sail, she was lively despite her weight, which much increased with soakage. I sold her only when we decided that however often we used a daysailer, it was less work to trundle one down from the shed and launch her than it was to keep one at a dock and bail her after every rain. Light weight then became crucial.

The hull of the mini-garvey weighs 55 pounds, and rig, board, and rudder add another 17. It would seem that she could easily be cartopped, but she does not accommodate two people graciously and one person would have a fair chore getting her up onto roof racks, so it would be better to launch her at sites where loafers are wont to congregate. With a trundle wheel in the mast partners she is easy enough to roll down to the dock, and there she can be picked up and launched sideways into the water, as is done with new million-ton grain barges.

Mini-garvey

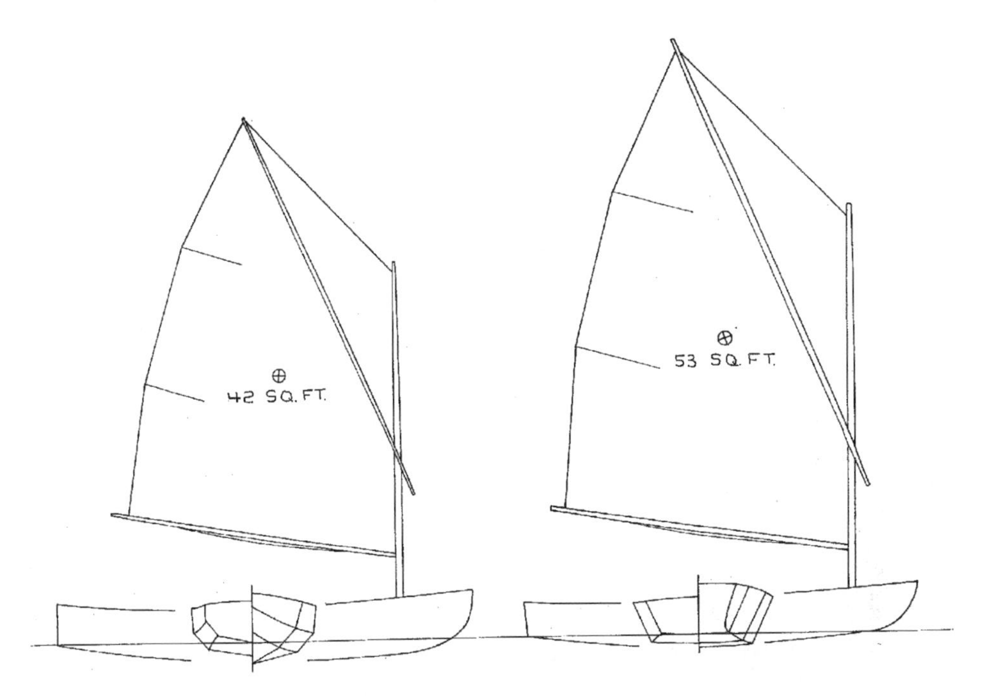


Lines and sailplans for the Tuckahoe Ten (left) and mini-garvey

The mast was meant to rotate, and as I said earlier, I would not repeat that experiment but would leave the mast square from the partners on down. The only other complication to the rig is a two-part vang that goes from the boom down to a block on the step and up again to a jam cleat. Unlike a racing vang, it can't be taken in when the sheet is eased and the sail is full, so it is set up loosely before leaving the dock and tightened before the sheet is eased. The boom is just low enough to graze the skipper's head when tacking if the vang is down hard.

The skipper sits facing forward on one of two cushions ("throwable devices," in Coast Guard parlance) set side-by-side behind the midships frame. His legs are on each side of the trunk, and as the yokes that support the trunk are less than 2 inches high, they aren't a bother. Tacking, the skipper slides over to the other cushion. The sheet comes down to a swiveling bull's eye and cam cleat on the aft end of the trunk. The boat is steered with a hiking stick, which requires a push-pull motion, to which not all sailors can accustom themselves.

In the Tuckahoe Ten the skipper sits in a similar posture, and sailing that boat I felt the need of a backrest made of canvas webbing. The garvey seems comfortable enough without one, however, bearing in mind that no daysailer is as comfortable as the worst chair in your livingroom. In strong wind you could get more performance out of either of these boats by sitting up on the deck and hiking out,

Spanish windlass on mini-garvey bows. Note doubling plywood for strip-plank fastenings

but our inclination then is to postpone the sailing until tomorrow. I do my rare bit of hiking in El Toro races, and Carol stays inside her boat no matter what. We are interested in the brain part of sailing more than the brawn part, although other sailors often like to exercise both.

The chines of the mini-garvey have only 3 inches of rocker on a bottom 8 feet, $2\,^1/_2$ inches by 2 feet, 6 inches, so even with the thwartships arc to stiffen the bottom planking, draft is less than 4 inches and displacement only 223 pounds. However, daysailers sail much better overloaded than underloaded. A V-bottomed 12-footer that I built many years ago for two hefty people sailed very badly when only one of them was in it, and the mini-garvey with both Carol and me aboard had no trouble pulling away from a competently handled Blue Jay, although it did not, of course, have the lively and responsive feeling that is the mark of a good daysailer. Carol is usually the one sailing the boat, so total weight is still less than designed displacement. I use it myself in winter when the water is cold because the Ten doesn't have much stability, being next thing to round-bottomed, and the garvey feels secure.

However, I do not like the bow noise. In the bigger carvel-planked garvey the cedar absorbed much of the noise, but the drumlike plywood mini-garvey sounds like a tidal wave in anything more than a zephyr. When she's reaching in fresh conditions that bow wave is impressive to see, and it's hard to believe that it isn't slowing the boat cripplingly. But when I'm in the Ten and Carol is in the mini-garvey, I may have my sail set to perfection and my boat may have a longer water-line and less wetted surface, but its pointed bow isn't cleaving the water as fast as the garvey bow is pushing it down. This garvey has too much rocker aft to plane, but when the wind freshens it does reach hull speed in a big hurry. In fact, acceleration is what people who have tried it most often remark upon: the mini-garvey feels something like a multihull. These people don't complain about the noisy bow. They may associate it with the rush that comes when the boat takes off, and certainly Carol does (although she can be a little deaf when she feels like it).

Building the mini-garvey was similar to building the *Dandy* dinghy: light framing on a strongback with light longitudinals and 5-millimeter okoume ply planking. The boat needn't stand as much abuse because it is not likely to be left untended for hours at a dinghy dock. Forward, a temporary spreader that is visible in the photo was put in to assure the arc in the bottom until the planking glue had hardened. The forwardmost frame is at the forward end of the bottom, and the shape of the bow was drawn and cut out on the topsides planking. Because that planking is not parallel to the lofting board, the shape is different enough to warrant the kind of lofting work most often associated with round-bottomed hulls, not hard-chine ones.

The bow of the topsides planking was then doubled with an extra piece of plywood to give some bearing for nails and glue and pulled together with a Spanish windlass, as can be seen in the photo. Spanish windlasses were often used in the making of planked garveys long ago, but how wonderful it is today to be able to drill holes for them wherever we like and fill the holes later with epoxy that is stronger than the wood and never comes out. The bow was then strip-planked with $^{3}/_{4}$ -inch-wide cedar strips the thickness of the bottom and sheathed with 6-ounce fiberglass outside and in. The nails through the strips were just to hold them in place until the glue set, and the glass came around the bow an inch or so, outside and in.

This is about as much strip planking as I like to do nowadays, although in the past I've built whole hulls this way. About an hour is needed to fit and glue up each square foot, and more time is invested ripping the stock and sanding and fairing once the glue is hard. The work is neither particularly pleasant nor imaginative. In the garvey bow I did use epoxy, although the strips are purely a core material for the glass skins, and I did bevel them to fit around the curve, although some builders would merely fill the interstices with thickened epoxy. Doing less than your best work is always less fun, and if the strips are beveled it is easier to see when one isn't lining up quite right and correct that with a staple or two. To complete four square feet of strip planking may have taken 8 hours, with glassing and sanding, and that's enough.

A sheer clamp is not needed in a light boat that has side decks, and it only adds weight. The topsides planking is allowed to overlap the marks on the frames, and then, before the hull is taken off the strongback, a guard rail of $^{1}/_{2}$ inches by $^{3}/_{4}$ inches is glued on flatways. After the hull is turned, the excess planking is planed off, and the guard and planking are beveled to accommodate the deck camber. This means that the deck plywood comes right out to the edge of the boat; end-grain ply does not take kindly to abrasion, but if it is well rounded over and epoxy coated, no harm should result.

The inner edge of the deck needs a coaming to span the 2 feet between deck beams, but $^1/_2$ inch square is enough. Too many builders use $^3/_4$ inch everywhere because they buy $^3/_4$-inch boards, and they hate to rip off a bit of what they paid for and waste it. But the extra wood all adds weight, and in this boat and the *Dandy* dinghy the frames are only $^1/_2$ inch thick. The round coaming at the bow was easier laminated than steamed because a steamed coaming would have needed a complete form in order to take a fair curve, while the laminated one spanned fairly around a few blocks temporarily tacked in place.

Mast weight is crucial in a light boat, and I have known boats to capsize at the dock with no sail set when the mast was too heavy. Once in Florida we sailed a friend's Laser that he didn't use very often but kindly rigged for us. He did not realize that the top half of the two-piece aluminum mast had filled with rain water since he last took the boat out. We did manage to have our sail and stay upright but came home with a very low opinion of the Laser; it was not until we tried one again several years later that we appreciated what a quick, handy, and pleasant boat it is.

For the Tuckahoe Ten I got away with an Engelmann spruce mast 1 $^3/_4$ inches in diameter. Multiplying the height of the center of effort above waterline by the sail area is the conventional way to calculate the load that a sailing rig puts on the mast and on the whole boat, and that suggested that the mini-garvey could do with a 2-inch Engelmann mast or 1 $^3/_4$-inch fir. But it didn't take into account the much greater stiffness of the garvey hull, and in the end I used 2-inch Douglas fir.

A mast of this strength would have to be 2 $^1/_2$ inches if hollow, and as the taper starts early and culminates in 1-inch diameter at the head, very careful hollowing out of the staves would be needed to yield a hollow mast lighter than the solid one. Like all spars (and rudders and daggerboards, too), a solid mast should be glued up from at least two sticks to guard against warping. In old-time boatbuilding books such as Chapelle we are told that a good mast can be made from a young tree, provided that the center of the tree is in the center of the mast. Such felled timber should be aged a year for every inch of thickness, and when used as a mast it inevitably checks, which causes anxiety if nothing else. And it still may warp.

Select Douglas fir from the lumberyard is less uniform since the spotted owl rulings closed much of the forest to logging. That's good for the owls, but it means we must do more selecting among the piles of selects. Much of the lumber is second growth from tree farms, so the annual rings are farther apart, and it often weighs more than the select fir we used to get. It's supposed to weigh 30 pounds a cubic foot, but through careless selecting I recently bought myself a board that weighs 43 pounds. Spars of that piece would have added several very undesirable pounds to the garvey rig.

As is, when we're getting under way and there's a decent wind blowing, I can stick in the Ten rig while sitting on the dock and holding the boat with my feet. The garvey rig is enough heavier (and with enough more bulk and windage, even brailed) that I must let go of the boat and stand up to put in the rig. But once you're in the mini-garvey and sailing, it is a wonderful boat—or it would be if the bow were quieter.

# Bobcat

Al Whitehead brought me his 15-foot plywood daysailer, of uncertain provenance, in the hope that I could mend it. In the six years since he'd acquired it, he'd worked on it a good deal himself, mostly with putty, glass, and paint. As it came without a rig, he had bought a new spritsail complete with very nice spruce spars from Roger Crawford, who customarily puts the rig into his fiberglass dories. The boat always leaked, Al said, and at first, he didn't mind the occasional bailing. But lately he'd had to bail almost continuously.

We looked her over, and I had to tell him that if there was a way to fix her I didn't know it. Amidships, the keelson had rotted away and disappeared, and Al had cut a crucial frame to get in there and make a new one with putty. Unluckily a trailer roller was right under that frame, and it had driven the centerboard trunk a couple of inches up through the deck. The boat had been nicely built and was nicely painted now, but under that glass and putty I was sure there was rot in the board trunk and the planking. Even if such a repair could be made, it couldn't be calculated or priced; and Al, who was middle-aged but had just gotten a new college degree and started a new career as a marine biologist, was understandably price-conscious.

We sat around the living room glumly discussing the options, and then finally Al went out to his car and brought in an issue of *Small Boat Journal* (a dozen years old, but apparently there is no limit to the life of good nautical magazines) containing an article on the Bobcat, Phil Bolger's plywood version of the Beetle Cat. Would his sprit rig work in this hull? And if so, and if he did the painting, what would the boat cost?

The Bobcat is one of Bolger's most popular "instant boat" designs. The Beetle Cat is a popular but curious one-design, in that the rules specify that it must be wood planked. The people who own the design (for long the Concordia Company) would not license other builders. *Spartina,* a novel that sold well a dozen years ago, showed how little its author knew about small sailboats when it said that the hero had "built a couple of Beetle Cats." Many good boats such as the Laser (and many very bad ones, too) are proprietary designs, and until recently the Olympic Committee refused to have them in competition, feeling that they gave to the country of manufacture and to rich countries in general an unfair advantage at the Olympic Games. Recently, with the world-wide triumph of wealth and wealthy self-righteousness, the Laser has been allowed in.

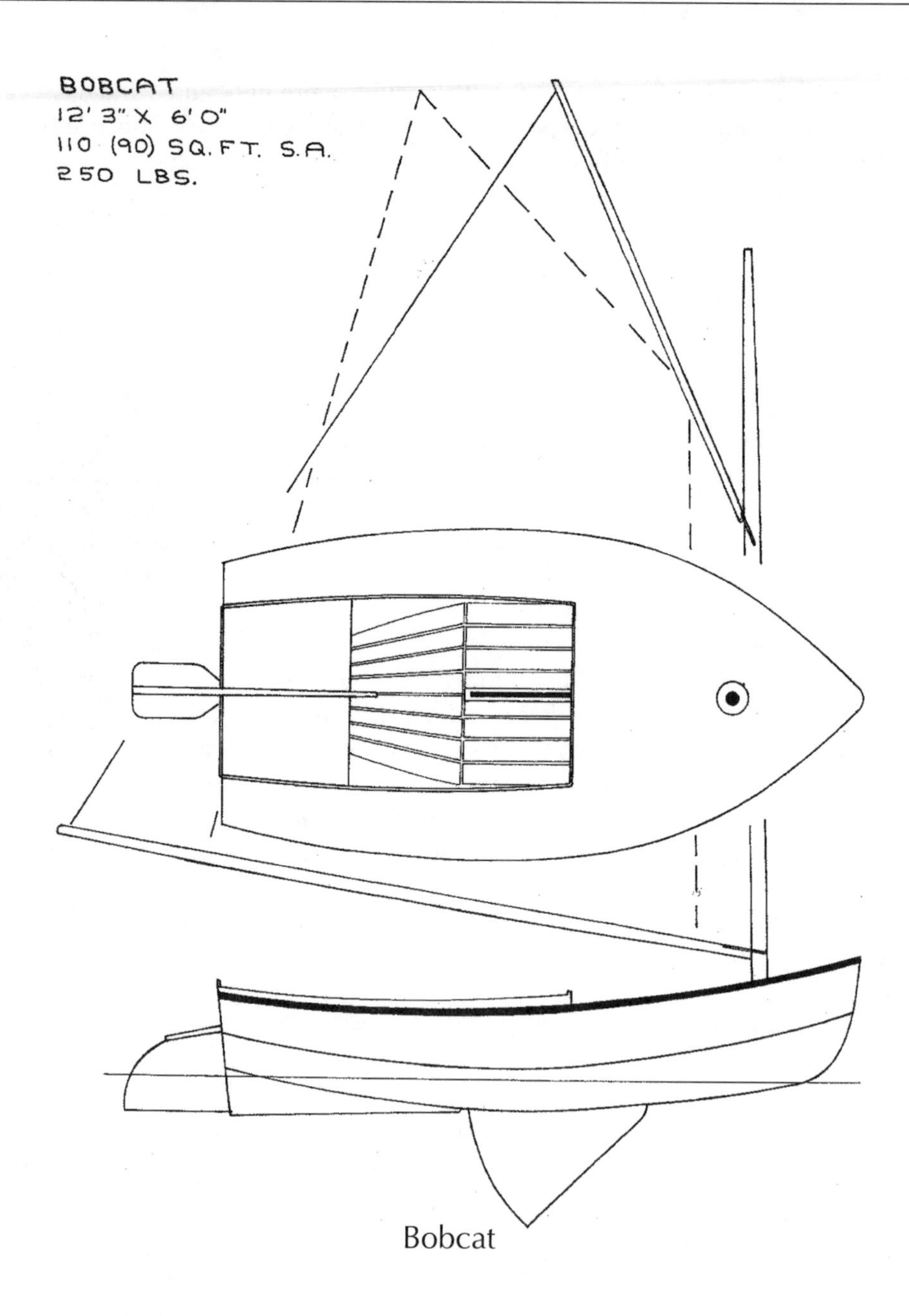


Bobcat

In his *SBJ* column Bolger called the Beetle Cat "a living fossil. It was designed in 1920, but was almost as reactionary then as it is now." Nevertheless, it has remained popular, and when we cruise to New England we often see fleets of Beetles used as trainers, although it is not certain how long yacht clubs will continue with them, with the competition from Optimists. Some years ago the building of Beetles was separated from Concordia, and Concordia itself is now shut. Twenty new Beetles a year was the production figure quoted twenty years ago, but it must be lower now. The price has gone up with inflation and now seems excessive to people who cheerfully spend $10,000 on a record player or $100 on a necktie.

French gig for the Atlantic Challenge

We've had a chance to study a Beetle over the years because a friend half a mile up the river bought a new one from Concordia a couple of years before we moved to Tuckahoe, and he still owns it. It has not been impressive: Nothing is bedded, not even the canvas that covers the planked deck, and a good many fastenings are mild steel and have bled. The boat has given more trouble and needed more repairs than you would expect from a well-built wooden boat. However, the hull has stayed tight.

Beetle Cats are carvel planked, and that's a method that I've stopped using, although it sure is fun to do. To make a tight hull it is necessary to use edge-grain virgin-forest planks to minimize shrinking and swelling. Even with that it's hard to keep the topsides of a sailboat free of leaks, because the boat sits on her mooring with the topsides exposed to a double ration of sun—both actual and reflected—and then is expected to be dry when she heels over. It used to be that people routinely bailed or pumped their boats, making little jokes about keeping the bilge sweet with the salt. When plywood, plastic, and aluminum came along, people came to see leaks as a sure sign of impending death, so a planked boat of today must be as tight as any other, and that's much easier done with clinker planking than carvel.

The last carvel boat that I helped build was when I was teaching boatbuilding to vocational technical students. It was a copy of a 38-foot-by-6-foot French navy gig, meant to compete in the Atlantic Challenge, the brainstorm of wooden boatbuilding teacher Lance Lee. He sought out a design that would be extreme: both hard to build and hard to use. The photo shows her about half planked. Since then, others beside myself have concluded that the French gig is not an ideal first-boat project, and the voc-tech students are now building instant boats. We did get the hull planked up before weather and the politics of the foundation I worked for drove the students away.

The planks were from Maine cedar saplings, which added to the challenge. It was not possible to spile them out without including a generous measure of sapwood, which rots fast, and the knots were legion and most of them loose. The idea was to build her with the materials of two centuries ago, but I did finally persuade the foundation to let us use epoxy glue when plugging the knotholes. Since then, the boat has been completed, although not by voc-tech students, and it's said that the hull is tight. There's no possibility that it will stay tight. However many corners Concordia was cutting on Beetle Cats twenty years ago, I'm sure they were ordering and picking over their planking stock with great care.

Bolger drew a flat-bottomed Bobcat with double-chine plywood hull, peaked the gaff more than the Beetle's, and added a dashing 10 square feet to the sailplan, no doubt based on his observation of Beetles. Al's sprit rig (shown with dashed lines in the drawing) is the height of Bolger's but narrower. I had to move the mast aft about a foot to keep the center of effort in the same place that Bolger has his.

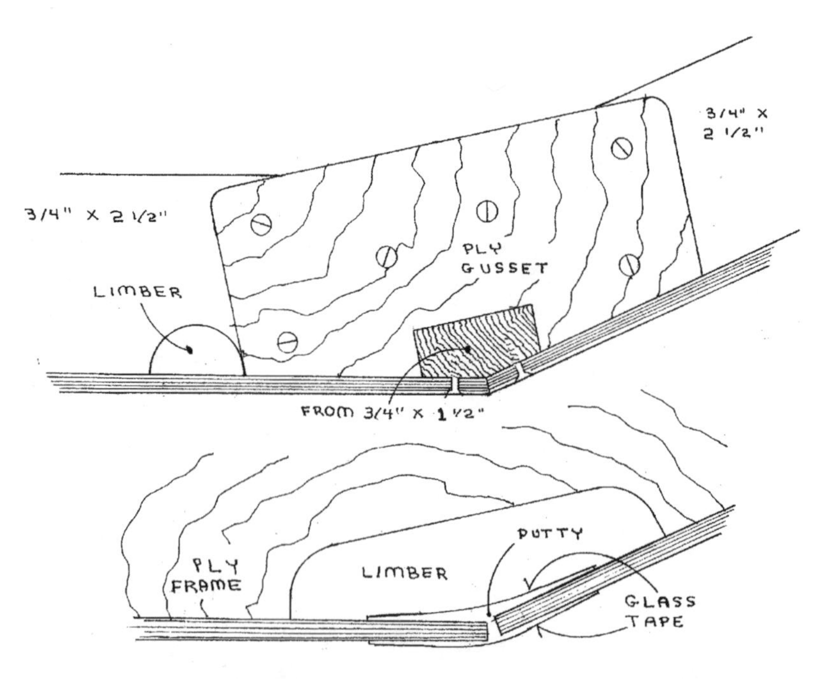


Taped-seam chine (below) and stringers-over-frames (above)

Modern thinking is that rig height is all that matters upwind, and total sail area becomes important only when the sheet is well out.

Phil was unhappy with this change to his Bobcat. Probably he thought that another professionally built example (Dynamite Payson had already built one) would yield some good photographs. The Whitehead boat is not stock, and probably Bolger finds it less pretty; but a builder makes these compromises to cut a deal, and who knows whether Al would have signed a contract if I had insisted on the Bolger rig?

I made a less objectionable change by putting back in the slight crown of the Beetle deck, which Bolger had flattened for building simplicity. A crown stiffens a deck, which is helpful if it is 1/4 -inch plywood, and cutting a few curved deck beams isn't much trouble. The curved cockpit coaming of the Beetle would have been considerable trouble, more in framing and fitting the deck than in steaming the coaming itself. Bolger drew a rectangular cockpit, and that's what I built. It gives a tiny bit more space inside this boat, although even that space is much constricted by the centerboard trunk. A daggerboard instead of centerboard would have made a far roomier boat, but the Beetle doesn't have one, and Bolger doesn't like them.

The biggest change was to build with stringers over sawn frames rather than the taped seam shown on the plans. The Bobcat is not a light boat (Bolger figures 250 pounds), and the sailing weight might be 450 pounds with skipper and gear aboard, not to mention possible crew. The extra 20 pounds that the wood framing added is not a great percentage of the whole. The 1/4 -inch plywood frames shown on the plans do clutter things, and they are subject to damage. They go straight down from coaming to sole, making a series of cubbies whose inboard edges are very likely to get kicked, and it might be wise to stiffen their inboard edges with a bit of lumber or, as Bolger suggests, use 1/2 -inch ply for the frames, but that would use up 20 pounds in a hurry. Ply frames also are expensive, as any grade of plywood costs about twice as much per cubic foot as the equivalent grade of lumber. When making frames, lumber also cuts to less waste than ply, despite Bolger's ingenious ply-cutting diagrams.

It isn't so easy to build by stitch and glue a double-chine hull half as wide as it is long. Payson says that temporary "edging" and "cleats" must be added to the plywood frames and taken out later, when some of the tape seaming is completed. Why isn't it easier to connect the edging and cleats to each other and dispense with the plywood in the frames? The bilge panels of this boat are the only ones with twist, and getting them into their proper places "requires patience," as Payson mildly puts it. On frames 2 feet apart you're trying to match up simultaneously the top and bottom edges of these floppy 13-foot sheets and to secure it all with wires. Bolger's offsets for panel cutting are exact, I'm sure, but the possible inaccuracy of your own cuts are what create the problems and call for the patience.

Dynamite Payson is an esteemed acquaintance of mine, and Phil Bolger is a good friend, so I hope none of this discussion is taken as criticism of them. They are simply trying to design and sell plans for an attractive boat with a building method that is currently in fashion. On the Bobcat plans Bolger provides a sectional lines drawing and a table of offsets, in case "substantial alterations from detailed design are intended." When I told him how I planned to build this hull he was pleased and interested, and Payson said it "sounds like real class" and asked to be sent a photo.

Why is it that taped seam has become so popular? It is the fear of beveling chines— especially if the bevels change. It is the fear of notching longitudinals into frames and of having to bevel off some of the frames later. But Glen-L's *Boatbuilding with Plywood* explains all this with ponderous clarity, and it isn't hard to do. For me, fitting two pieces of wood together as accurately as I can is the fun and challenge of wooden boatbuilding.

A few tips may be useful for stringers over frames. The screws that attach the stringers to the frames should be well sunk in; your plane will not improve from hitting the head of such a screw, especially if it is stainless steel. The hole can be filled with epoxy later, if necessary. While beveling I use a leg attached to the plane, with a shoe that is exactly parallel with the plane bed. Then, by running the shoe along the longitudinal next above or below the chine log being planed, a good bevel is assured. It is true that forward on the bilge panel of this boat there is quite a little sectional curve due to the twisting of the sheet, and the shoe and leg won't work perfectly. A little eye-balling is needed. But by starting aft where there's no twist, you gain enough confidence by the time you reach the bow. Where both frame and chine log must be beveled, it's best to take them down alternately, a little at a time.

When planking, trips must be made underneath the hull to scrape out excess glue that squashes out when the screws are taken up tight. It is tempting to use this excess epoxy to fill screw heads because otherwise it must be thrown away. However, epoxy thickened with silica cures so hard that in sanding the wood around it is likely to become dished. It's best to wait to fill the screw heads until the whole hull is planked and then do it with a less dense thickener than silica, such as microballoons.

Because of her great beam, the Bobcat is not easy to frame. To make the upper chine logs take the bend, they have to be put in on a tilt so that they are not parallel to either the topsides or the bilge panel, and then both edges have to be beveled. With a longer, narrower boat—such as a multihull—chine logs can always be put in parallel to one surface, and only the other surface need be beveled. Even by tilting, the lower chine log cannot be persuaded to come to the stem, and a small wood block has to be added there and the whole thing rounded off after the glue has set. That must make a slight difference in the shape of the bow, but I doubt it's disastrous, and the ply goes around it fine.

On single-chine plywood designs, books tell us to use an overlapping join in the ply skin aft and switch to a miter join toward the bow, where the overlap would become too wide. On this double-chine hull I used miter joins in all seams from bow to stern. The more chines a hull has, the more obtuse their angles and the less point there is in overlapping joins. For the bottom panel, which was the first to be scribed in place, I set the Skilsaw at 30 degrees; I did the same for the bilge panels. Then, with the bottom screwed and glued in place, the bilge panels were offered up and their bevels perfected with a block plane. This made a very neat job and elimi-

Bobcat interior; note intermediate stringers in bilge panels

nated ply end grain from the whole hull surface except at the transom. Butts were made on the boat, so I never had to handle a ply plank over 8 feet long, as must be done when tape seaming. And by putting on the bottom plank first, then the bilges, and then the topsides, I never had to be fitting two edges at once, or putting in a shutter, as is said in carvel building.

The drawing shows a place where it is nearly impossible to make a neat job of taped-seam construction. The tape on the inside of the chines is supposed to go right through a limber in each ply frame. Imagine threading glass tape through there! Imagine sanding it after it has two coats of epoxy resin! Even though using chine logs instead of glass tape requires a separate limber hole, it looks pretty clean by comparison and pretty easy to do. Small taped-seam boats such as kayaks usually do not need frames, but with larger ones that do, the inside is likely to be a mess.

Perhaps the greatest long-term benefit of building a Bobcat with stringers is that it allowed me to put an intermediate stringer in the bilge panels, visible in the photo. Bolger admits that many of his taped-seam designs, if used hard, eventually become very flexible, and the large unsupported panels of thin ply are part of the reason. The chines of the Bobcat are 2 feet apart in places, and the frames are also 2 feet apart, and that is too big an area to be spanned by $^1/_4$-inch plywood. Like the lower chine log, the intermediate bilge stringer could not be bent enough to touch the stem, but up there the panel has enough compound curve to be stiff.

Al Whitehead's Bobcat

One bit of fun building a Bobcat is putting the slug of lead into the centerboard. All boards, center- or dagger-, tend to float up when fully down unless made of very heavy exotic wood. Even the white oak centerboard in the 90-foot Delaware Bay oyster schooner that friends of ours are now restoring would float up if it didn't have many hundreds of pounds of iron drift bolts, which are also there to hold the massive oak chunks together. A small daggerboard can be held where you want it with a bit of shockcord, but for this centerboard Bolger calls for 10 pounds of lead. It's 6 inches square, and the board is $^3/_4$ inch thick. Some builders use lead sheets surrounded by epoxy, but lead is easy to heat and pour in small quantities, and working with it is a pleasant break from the epoxy routine.

The hole is cut in the board with the Skilsaw set at 45 degrees, first on one side and then the other, to give a sharply veed edge in the middle layer of the plywood, which helps keep the lead from falling out some time in the future. To secure it even better, bronze nails or screws can be driven into the point of the V, to be surrounded by the hot lead. Bronze and lead are very close on the galvanic scale and do not make trouble for each other.

To back up one side of the hole in the centerboard, I clamped on a sheet of $^1/_4$-inch aluminum, which, like copper, absorbs heat fast and so minimizes charring of the wood around the hole. Lead melts at 558 degrees Fahrenheit, which is far below the melting point of aluminum but hot enough to damage wood. The board, of course, must be leveled up near the stove, and then the fun begins.

I used a couple of chunks of lead from a plumbing supply house, and I wore leather gloves and shoes and loose-fitting clothes. One thinks, good Lord, this is going to need a lot of heat! Five hundred fifty-eight degrees is two and a half times the boiling point of water! But that's just a trick that our measurement system plays on us, and the little propane hotplate in my shop, which produces far less heat than a burner on a kitchen range, melted this lead almost as quickly as it boils water. Once the lead is poured and has cooled, the excess can be taken off with any woodworking tool, such as sandpaper or a block plane. You should wash your hands afterwards, and you should take care not to ingest any lead dust. They say that the Roman Empire fell because people got goofy from drinking water that came through lead pipes, so as with epoxy and many other boatbuilding materials, we need to be careful with lead.

Al Whitehead's contract specified that he was to do the painting, which suited me fine. He named her *Mayfly,* which is a very different creature from a bobcat but seems to suit her nonetheless. He brought her down for trials and photos and then again about a year later so that we could have a day together on the river. I had a chance to sail the boat and to compare her with the mini-garvey, which was also launched that day.

The space in the Bobcat is limited, but both her beam and weight make her so stable that you don't have to shift position in normal winds. A second person aboard really needs to sit up beside the board trunk for fore-and-aft trim; the big back seat that looks so inviting is best avoided. The seat is there to make a level surface to walk on because the hull rockers up abruptly in the last few feet. I have seen Leo Telesmanik, who ran the Concordia Beetle Cat shop for decades, racing a Beetle with another old man, and they sat on the bottom just aft of the trunk, facing each other. They traded helm when they tacked and never moved.

Compared with the very light sailboats that we are used to, the Bobcat gets under way rather deliberately, and in all her responses she feels more like a cruising boat than a daysailer. Perhaps the extra sail area that Bolger drew would make her livelier, but the rudder also contributes to this slow response. Under water it is 8 inches deep and 24 inches long. As the drawing shows, Bolger has an endplate on it a foot wide, which is supposed to increase efficiency, but it feels mighty heavy in the hand, and the boat turns slowly. There are nearly 7 square feet of wetted surface in this rudder—and then there's the skeg. The garvey, which has no skeg and a rudder 15 inches deep and 7 inches long, turns like lightning. Catboat rudders like the Bobcat's don't impress me, when a kick-up rudder would give the same access to shallow water.

The Bobcat does truck right along, for all that, and although she tacks slowly, she never fails to tack. She balances well when sailing level, but when heeled she develops strong weather helm; if she heels enough, of course, she picks her rudder out of the water and luffs up. In designing her, Phil hoped she'd sail as well as a Beetle Cat, but when his wife, Susanne Altenburger, finally got him to sail one (it's hard to pull Phil away from the drafting board), he found her superior in every way. That seems likely because, if nothing else, she weighs 150 pounds less. Her upper chines are well above the water, so her waterline beam at rest is only 4 $^1/_2$ feet, and it is often less under way, while the Beetle must be over 5 feet wide in the water at all times.

For Al Whitehead the great revelation of *Mayfly* is that she doesn't leak. The first year he sailed her, in every phone call when I asked how she was, he replied, "Great! She doesn't leak!" Apparently he had put up with that leaky old hull for so long that he only hoped for slower leakage in the new one and never dreamed that it could be eliminated.

# Fiberglass Dobler

It was on a vacation in the Caribbean that the Mahlers decided they wanted a sailboat. With another couple they chartered a big motorsailer and three crew. Like everything else about the trip, the sailing was super, although they weren't sure whether or not the motor was running during the sailing. When they called me up to inquire about a boat, I pointed out that they'd have to learn to sail. "Oh we will, we will," said Gail. "In June when the yacht club gives a sailing course I'll sign up."

The Mahlers are good people. For years Gail volunteered at the Avian Rehabilitation Center, where birds were sometimes nursed back to health. Living on the water and having a dock, they naturally owned a runabout, but Keith wasn't interested in fishing, let alone the dumber runabout entertainments like waterskiing, so for years he motored up and down on weekends picking up debris: Spackle buck-

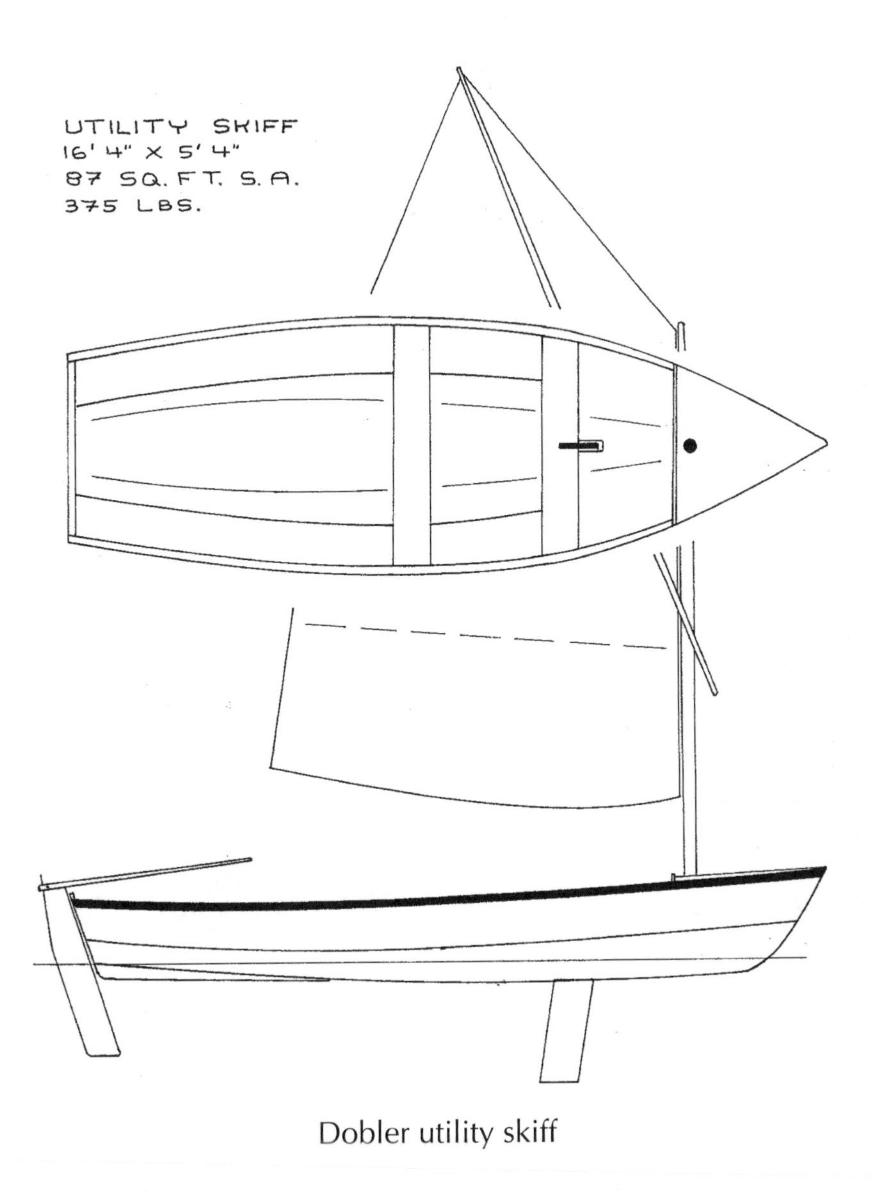


Dobler utility skiff

ets, boards full of nails, Styrofoam billets from old docks. He is a big man, and he wanted his daysailer to be big and heavy for initial stability when he boarded it. Because he expected to keep the boat for many years, he also wanted the hull to be fiberglass.

I'm not sure that fiberglass boats last longer than wooden ones. They say you don't have to maintain them, so people don't. The gelcoat starts sloughing off after about ten years. The upholstery (why do they always have upholstery?) starts disintegrating even before that. The no-maintenance teak trim looks like driftwood after a few seasons. The Plexiglas cracks or clouds over, the canvas comes unsewn,

Fiberglass Dobler in very light air

and the Zamak hardware puts out as many bubbles as a dying crab. All of these problems can be fixed or replaced, but none are because the whole idea of a fiberglass boat is that you don't maintain it. So eventually you don't even want to step in it, let alone touch it. And the design is obsolete now: it doesn't look like the ones in the magazines any more.

There is a market for astonishingly old outboard motors. You see them advertised, and presumably some are sold. On the Maurice River there's a marina where nobody ever takes out a boat. Summer weekend skippers head down to it with six-packs to talk with their buddies and work on their crummy old outboards. But there is no market for crummy old fiberglass boats. When Carol and I first moved to Tuckahoe twenty years ago, the boats that came drifting up with the tide were wooden, and they were well and truly derelict. These days they are usually fiberglass, and they are just as truly derelict. They don't biodegrade, but the marsh still seems to swallow them. A winter storm tide takes them well up into it, and next spring the phragmites grow up around them, and in a year you can't see them or remember where they are. They may provide some habitat for wildlife, as birdhouses are alleged to do. I suppose the marsh will eventually be saturated with them.

There are several ways to build a one-off fiberglass boat. You can set up a male mold with wooden frames and closely spaced stringers. Foam sheets about 4 feet square in various thicknesses and densities are then sewn to the stringers or screwed to them from inside, and the outside is glassed. Then the threads are cut or the screws are extracted, and the hull is taken off the mold, turned over, and glassed on the inside. A foam sandwich hull is very stiff and considerably lighter weight than a solid fiberglass hull, with some thermal and noise insulation. However, the Mahlers didn't want a light hull, and the foam itself is about as expensive as the best plywood, before adding the cost of fiberglass and resin. You must finish with a sander and with gelcoat or paint both the outside and inside, and finishing the inside of a fiberglass hull is especially nasty work, with the glass dust swirling all around you.

C-Flex has been sold for years for the making of one-off fiberglass hulls, and I did seriously consider it for the Mahlers' daysailer. It looks like fiberglass fishing rods held together by light scrim and comes in rolls a foot wide. Molds are set up, and the C-Flex is stapled to them. It should then be saturated with resin, although I have recently seen a hull where this step lamentably was omitted. Then layers of mat and roving or cloth are put on until the desired thickness, strength, and stiffness are achieved.

One problem with C-Flex is that the resultant hull interior is very rough. It needs some grinding, a layer of mat, and then some more grinding. Often such hulls are completely lined with wood or plywood, which suggests that C-Flex is better suited to a cruising boat than a daysailer. The wooden mold over which it was originally stapled does not come out easy (especially such bits as the stem), and often builders leave some parts of the mold in place and perhaps glass over them.

The Mahlers' boat was built by laying up sheets of fiberglass on a table, cutting them to shape, and tape seaming them together. This method works only with a chined hull, but I had already convinced myself and my customers that what they needed was a 16-foot, 4-inch utility skiff designed by Joseph Dobler. The boat was designed for plywood construction, and for several years Carol and I owned one. It may be one of the best daysailers ever drawn: not too hard to build, simple but sprightly to sail, phenomenally comfortable for up to seven people (although it does slow down with more than two aboard), and adequate to row when the wind grows fickle.

Dobler derived the design from Swampscott dories, but it is a better boat for the same reasons that a Bobcat is better than a Beetle Cat: the flat bottom and double chine make it narrower on the waterline but give it more reserve buoyancy; and it is lighter, with less soakage in the plywood than there is in boards. With a Sunfish rig, ours outsailed any Sunfish except close hauled, where the greater freeboard and windage slowed her down.

A comparison of the Bobcat and Dobler drawings shows that the Dobler has an easier-driven hull and can get away with less sail area, though it is 4 feet longer. In discussing the Bobcat, Bolger himself displays considerable impatience, saying, "The Beetle Cat would be more boat for the money if it were stretched out three feet on the same breadth and depth." By implication that applies to all short, fat catboats, but Bolger concedes that lots of people—including the *SBJ* reader who asked for the design—do love catboats.

Our Dobler we sailed in Cape May's big harbor, studying up the fleet that came through on the Intracoastal Waterway, which was more varied and interesting than it has since become. I wrote about her in my first boating book and am embarrassed to say that she generated more inquiries than any design of my own.

The hull of that boat weighed 215 pounds, seats, deck, and all, but Keith had in mind 400 pounds, so there was no point in a sophisticated layup. Many very heavy but durable and good mass-produced boats have been built with pure mat, out of a chopper gun. When the Mahlers' hull was built polyester resin cost 95 cents a pound and mat cost $1.50. Cloth was $2.50 and woven roving about the same. It is easy to see that for economy resin is the material of choice, and it has enough strength by itself, but it needs glass for stiffness. Carefully laid mat yields a glass-to-resin ratio of 1:3. I bought and used up a 100-pound roll of 2-ounce glass mat in this skiff.

At one time Vance Buhler was employing nearly fifty people to build huge catamarans for tourist headboats on the Caribbean island of St. Vincent, and I was lucky enough to learn some of his many tricks. He used a Formica-covered table for the layup of many parts, such as the fronts of food-and-drink service bars, and would put a bit of foam or balsa core into the sections that were to remain flat and lay up just the right amount of glass in the sections that were to be radiused 90 degrees, so that they would bend when taken off the table; more glass could then be added to get the desired stiffness. The outside was, of course, the side that had been laid against the Formica, so it came off the table with a perfect finish.

Instead of Formica, I made my table from two sheets of melamine-covered Masonite. It worked fairly well, but after it had been used a couple of times it became less eager to release the cured sheets of fiberglass. That problem was resolved by painting it with a coat of mold-release wax before each layup. I doubt that using a plywood table would be a good idea because the surface is porous, so many coats of wax would be needed before you could be sure that the resin wouldn't stick to the table, and most plywood has some irregularities in the surface, which would transfer to the fiberglass.

Joe Dobler is said to be the first designer to bring the taped-seam method to America. It may have originated in England with Jack Holt's Mirror dinghy. But Dobler's method does not show us the panel shapes to be cut out, the way *Chingadera* does. Instead, we are given the dimensions of frames, and each panel has an intermediate stringer notched into the frames, although, of course, they need no beveling. The panel shapes are found by laying sheets of ply on the framework and marking them at the chines. Only the chines are tape-seamed.

I needed to know the exact shapes of my fiberglass planking panels from the start because offcuts from big sheets of fiberglass are less likely to be useful later than are plywood offcuts. Anyway, I wasn't going to use permanent frames or intermediate stringers as Dobler does. I made a quarter-size half model of the hull, planked it with $^1/_8$-inch plywood, and from the plywood took offsets. These offsets were multiplied by four and the shape marked out full-size on the table with a felt-tipped pen.

The planking panels were laid up, followed by the seat tops and risers, the transom, the small forward bulkhead, and the foredeck. The deck was given some camber in the Buhler manner, by laying it up thin on the table, springing some curve into it when it cured, and adding more glass. The hardest job was making the daggerboard trunk: glass was wrapped around a wooden plug covered with waxed paper. Talk about messy! The plug needed to be several pieces, with tapers, so that it could be extracted after the glass cured.

Two layers of mat seemed enough for the bilge and topsides panels, but the bottom is wider with little curve to add stiffness, so three layers seemed safer there. When the panels were ready, I made a kind of ramshackle temporary framework to hold them in place. The panels were laid on the framework with the smooth side from the table inside and were screwed down here and there. When necessary they were aligned with duct tape. They were then "tacked" together with small squares of glass cloth. When that cured the screws and duct tape could be removed, and all seams could be filled with putty and taped with cloth outside and in. With polyester resin the putty need not be catalyzed because the hardener in the wetted tape sets it off, and you have more working time. After all that was done, I put two more layers of mat over the whole boat.

Anyone building even the simplest fiberglass boat should have a 7-inch high-speed disc sander, a good respirator (not just a mask), and either plenty of clothes that are nearly ready to throw away or several paper suits. I take care that I don't build in fiberglass more than once every couple of years, because it's easy to burn out doing it, easy to start drinking too much after the day's work is over, and then easy to begin drinking before the day's work begins. Anyone who has worked in a boatyard knows the habits and mindset of the confirmed fiberglass man.

Even one boat can drive you to drink if you decide, as unimaginative people often do, that the goal is cosmetic perfection. With fiberglass you can waste even more time this way than with a wooden boat, and because fiberglass is thinner you can even compromise the strength of a hull by sanding too much. The cheapest mass-produced fiberglass boats are often perfect on the outside. So is the coffee mug that you buy at the Swillmart for 59 cents. The coffee mug that you buy at the craft fair for $15 is often remarkably imperfect, and that's one of the reasons you buy it. Part of what you're paying for, beside function and an attractive design, is the evidence that this mug is one-of-a-kind and was made by human hands, not by a machine.

It's at least as true that handmade boats shouldn't try to look as if they were machine-made. With fiberglass you must sand until the skipper can't cut himself on anything. You must fill some gouges, but not every ripple. It's a question of knowing when to stop. You should eyeball the sheer before cutting it, but don't start eyeballing the planking for fairness. Especially don't buy one of those long sanding boards.

Despite the annoying sanding, there's something fascinating about fiberglass work. Often so much can be done so quickly and easily. The preparation takes more time than the layup. The pieces of glass need to be cut, marked with a felt pen, and stacked sequentially. All tools and supplies need to be arranged where they can be easily seen and picked up. Because you won't have much time when the resin is catalyzed, you have to be sure there won't be hitches or delays. But then you don the gloves and get going, and it's amazing what can be accomplished in a half-hour. Bingo!

Seats for the Dobler skiff were put together on the table using some right-angle brackets, clamps, and duct tape to hold them together and aligned on the outside while they were glass taped on the inside. Then the outsides of their corners could be ground, filled, taped, and ground again. They were scribed to the hull and held in place with T-brackets and sticks until they could be taped into place, as the photograph shows.

Gelcoat can be brushed on, but it doesn't achieve much. Production builders still use it because it can be sprayed into a mold quickly, but it's a poor moisture barrier and only fairly good at resisting sunlight. Nowadays the better production builders put on epoxy coatings to guard against osmosis. For a one-off it's better to put epoxy resin (if needed, and it isn't on boats stored out of the water) straight over the layup, followed by paint. Latex is easiest to apply, clean up, or touch up; but the Mahlers wanted a high gloss, so I brushed on one-pot polyurethane. It's not too different chemically from alkyd, but on this boat it lasted five years in Tuckahoe, never out of the sun, summer or winter. Now it's been in Florida a year, and Keith says it needs work. The last alkyd paint that I used on a boat—and very expensive alkyd paint it was—clouded badly in less than two years. Despite what the can says, brush-on polyurethane can be thinned with mineral spirits and cleaned up with that or with kerosene.

Varnish lasts a shorter time, and the Mahlers insisted that everything wood on the boat be varnished: thwarts, guards, spars, tiller, and rudder. Until they moved the boat to Florida, Gail did keep up with it. With varnish even more than with paint, only the very best should be used, because the material cost is so small compared to the labor. I get almost two years out of the best varnish on horizontal surfaces. Urethane varnish is harder and better for abrasion, but oil varnish holds up better in sun.

Fiberglass seats glassed in place

I added polymeric nonskid to the last coat of paint on the sole. It provides just the right degree of roughness and isn't sharp, so it's fine for bare feet or knuckles. Once mixed it doesn't settle out the way sand does or float up the way crushed walnut shells do because it's made to be exactly the same weight as paint.

Trimming out a fiberglass hull is like trimming out a wooden one. All wood should be glued or bedded. Many production builders don't bother, and they tell you that teak needs no bedding or gluing (or finish coatings, either). Bedding is the forgotten material in modern boatbuilding, and some of the old-timers didn't bother with it, either. The State of Maryland once had a repliboat, the square-rigged *Maryland Dove,* built by Jim Richardson, who was at that time the dean of the Chesapeake boatbuilders and now, being dead, has a whole museum devoted to his work. In not one place where wood touched wood or metal was there any bedding. After seven years of service, the *Dove* was hauled for "a little work," which wound up costing half the original building price. After the repairs, a friend who had worked with me at Yank's was hired as full-time bosun to keep up with the maintenance, mostly paint and bedding.

On the Mahlers' Dobler I added 12 square feet to the most successful of the several rigs we had had on our plywood Dobler, which was, of course, a spritsail. I also moved the rig and board trunk a foot forward, to get them farther out of the Mahlers' way. Traditional wisdom says this move unbalances the sailplan, because the rig's center of effort must be a certain percentage of the waterline length forward of the entire hull's center of resistance. That may be true for boats with long keels, but for modern boats with fin keels or daggerboards the center of effort should be directly over the keel or board, and the rest of the underwater shape can be disregarded.

The Mahlers were excited about their sailboat, and while I was building her Keith was building davits on his dock so that she could be hauled up away from wakes; I put a plug in the transom that could be pulled to let rainwater out, so that she would never need bailing. I bought them a primer on sailing to read during the winter, until Gail could take the yacht club course, and found that they had already bought themselves one and were much absorbed in it. Some people may be able to learn sailing or airplane flying entirely out of a book, but most of us can't visualize well enough from the written word, and we need an instructor. So it proved with the Mahlers. Gail went down to sign up for the course but found she was too late, and so far as I know neither of them has made another effort. From the books they know something about sailing but not enough to make it fun, so they sailed very rarely on the Tuckahoe. Their Florida house is on a canal two miles from the Caloosahatchee River, so it's not likely to be sailed more often down there.

I was out in the boat a couple of times, but the wind was so light that I didn't learn much about it. The big comfortable space with so many places to sit or stretch out was as delightful as it had been in the plywood version. Naturally a builder would like to see his product used and enjoyed, and the Mahlers may yet take a sailing course and get the hang of it. Or they may sell her and a second owner may use and enjoy her. In the twenty years that I've been building, that has happened to my boats a number of times.

# Melonseed

Fifteen years ago John Guidera decided it was time to buy a sailboat. He already knew how to sail, but he wanted to do it more often and to cruise a bit. Like many married men, he hoped to interest his wife in sailing, too.

He bought a used 22-foot trailersailer, which he kept in a marina but brought home winters. For years he and Gloria did sail together, but beating across Chesapeake Bay under storm canvas and (when the motor wouldn't start) beating into crowded Annapolis convinced her that she really didn't like it (not everybody does). John was now a singlehander.

Too many husbands try to whine, cajole, or threaten their wives into sailing, but I don't think John ever did, and I can't think of a worse wife to try those tactics on than Gloria. He found that the Tanzer wasn't much fun to sail solo, and my own limited experience of singlehanding a cruising boat was the same. Roger Taylor,

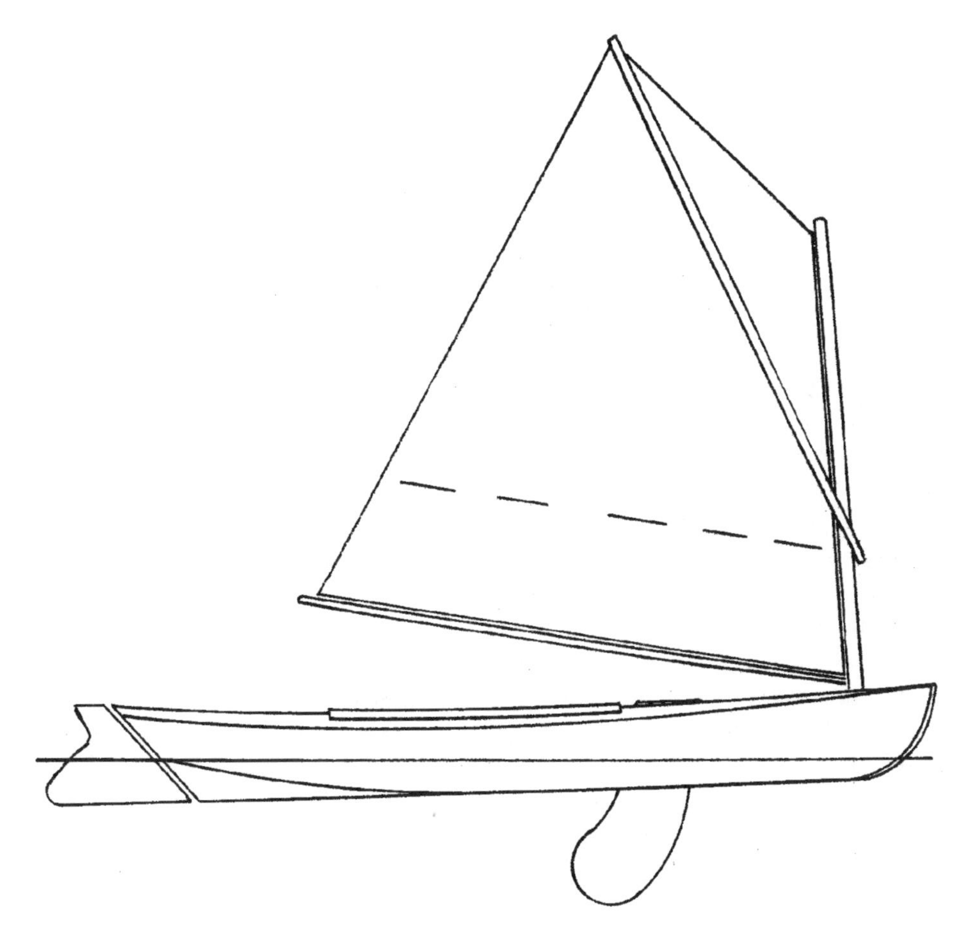

Working Melonseed after Chapelle

the former publisher of the *Small Boat Journal,* says that the chief purpose of sailing is the observation of the weather, but that loses punch when there's no one to discuss it with. There is also the hassle of managing several sails while reading the chart and keeping one eye on the compass, and the long dead evenings on anchor. After ten years John sold the trailersailer.

He decided he wanted a daysailer that he could readily manage by himself, which meant one sail. He wanted to trailer it and launch it from various sites so he could sail on various waters, which brings some of the pleasures of cruising. He wanted the boat to sail well and to be distinctive.

At the Philadelphia Maritime Museum he saw his first Melonseed, and later he read about the type and then sailed a fiberglass one with the builder, Roger Crawford. Crawford was agreeable, as all salesmen who make a living must be, but the wind was too light for John to draw conclusions. Besides, he remembered the Philadelphia boat and thought he'd rather have a wooden one if he could afford it. He planned to keep it in a garage, so rot and maintenance were not issues, and to his mind the idea of a traditional fiberglass Melonseed was ludicrous.

Among the hundred traditional small workboat designs that Howard Chapelle gathered together for his wonderful book, *American Small Sailing Craft,* the Melonseed is currently very popular. But building technology has improved enormously in the fifty years since Chapelle wrote, and traditional materials have become scarcer and not as good. In addition, an ideal workboat that looks after herself while the owner gets on with the work does not necessarily make a good pleasure boat, in which the owner can concentrate fully on making her go.

The Melonseed was developed to be a more seaworthy Sneak Box for New Jersey market gunners. It was rowed much of the time and perhaps sculled for the last few hundred yards, when sneaking up on sleeping ducks. A shotgun of perhaps 1 $^1/_2$-inch bore might have been bolted to the foredeck, and the skipper might have set it off by pulling a string from aft. The boat wasn't meant to be sporty. It had much deck and a tiny cockpit, a bare 6 inches of freeboard, and rails along the sheer to hold decoys. There was a sailing rig of 53 square feet, which was easy to handle and could stow inside the boat. A rudder that bit only 6 inches of water was controlled by a yoke and ropes.

Comparing the drawing of Chapelle's Melonseed with my plans shows the changes required. To figure out the hull shape for a pleasure version, it was first necessary to redraw Chapelle's lines, because he recorded them as he took them off the hull, and the crucial numbers—displacement, center of buoyancy, prismatic coefficient—cannot be figured from his work. For a job like this, and for any but the simplest hull drawing, it helps to draw the station grid and the waterlines in profile and buttock lines in plan—in other words all the straight lines—on one side of the tracing paper and the curved lines on the other, because in erasing the curves (as must be done often) one does not erase the grid.

Again using both sides of the paper, I drew a new set of lines with less deadrise to give less displacement and fuller ends to suit the lighter load and lighter weight of modern materials and the bigger rig, which would drive the boat faster. The center of buoyancy was moved aft several inches to better suit tiller steering. Freeboard was raised 2 inches, but when the molds were set up and John came over to look at them, we quickly agreed to raise it another 2 inches. It would have improved the boat in every way if the transom rake had been reduced, as has been done by Lyle Hess in what he persists in calling Bristol Channel Pilot Cutters. But the 45-degree transom (probably the extra deck aft was wanted for more decoy stowage) is one of the distinctive marks of the Melonseed, and John wanted it.

With such a raked transom the forward end of the tiller must be high, or it hits the coaming when put over. I got rid of the scimitar daggerboard, which requires a long trunk in which it tends to waggle around. The straight board cost 6 inches of cockpit space, but it's still 4 feet, 3 inches long, which is enough for a singlehander. Crawford has eliminated the decoy rails from his Melonseeds, but John wanted them retained. Rudder depth was increased to a daring 1 foot, and the skeg was shortened to reduce wetted surface. It pierces the bottom plank and is glued and screwed to the transom, making it as crash-proof as anything in the boat. While a

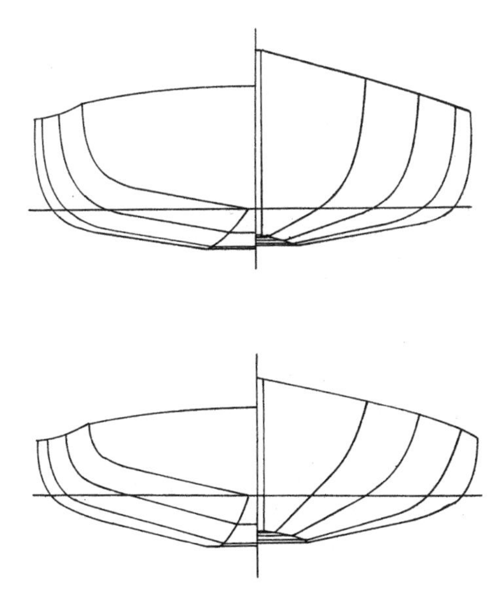

Melonseed sections: Guidera (above) and Chapelle (below)

market gunner might want to navigate very shallow water where ducks often sleep, a pleasure sailor has less reason to.

I'd have preferred a boom, perhaps a sprit boom like *Mayfly*'s, but John wanted a boomless sail, at least for starters. It's one less thing to rig when launching, and to adjust the tension under way it probably would have been necessary to crawl out on the foredeck. As we have all discovered, the sheet lead of a boomless jib must be taken from 40 percent up the luff through the clew and to a block or cleat; if taken from 45 percent up, or 35 percent, the sail sets very badly. A boomless spritsail seems more tolerant of sheet lead angle, and a lead anywhere from 65 percent to 80 percent up the luff works okay.

Working Melonseeds were built carvel or clinker, but John agreed that glued-lap clinker would be the best method to guarantee a tight hull in a wooden boat that would spend much of its life in dry storage. Fifteen years ago I build a clinker cedar kayak with polyurethane stickum in the laps and clench-nail fastening. It was always in a shed, and for ten years it didn't leak a drop, but it does now, and it's not going to be fun to fix. The second-growth lumber comes and goes, and the nails might put up with that, but the polyurethane has hardened over the years, and the planks have split themselves in a couple of places. But glued-lap clinker is as dimensionally stable as any plywood boat and lasts a very long time if moisture-attracting dirt is kept from accumulating in the laps.

I had never tried a round-bottomed hull in glued-lap and was looking forward to it. Tom Hill's *Ultralight Boatbuilding* is the chief reference on the subject. His method is to put a mold stringer at each lap. However, he admits that it is a

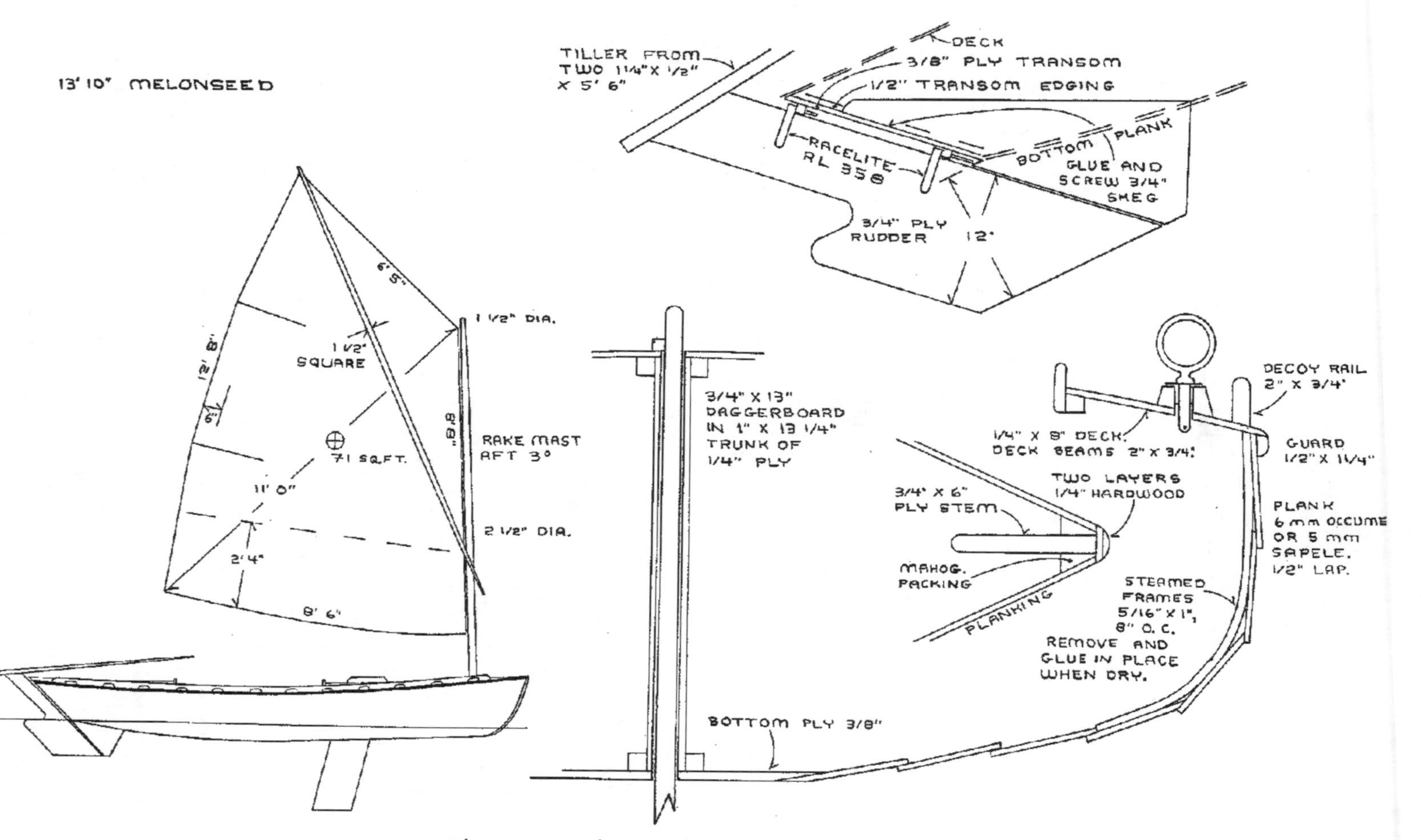


Pleasure Melonseed profile and details

MELONSEED OFFSETS, INCHES AND EIGHTHS

| STA | TO BASE SHEER | 18" BUT | 12" BUT | 6" BUT | BOT. | HALF BREDTHS SHEER | 16" WL | 18" WL | 20" WL | 22" WL | BOT. |
|---|---|---|---|---|---|---|---|---|---|---|---|
| STEM | 2-0 | — | — | — | 21-2 | 0-6 ← | | | | → | 0-6 |
| 0 | 3-0 | — | — | — | — | 4-3 | 2-2 | 1-3 | 0-6 | — | — |
| 1 | 5-2 | — | 11-6 | 21-0 | 23-4 | 12-5 | 10-6 | 9-5 | 7-3 | 3-6 | 1-3 |
| 2 | 6-6 | 13-0 | 21-0 | 23-7 | 23-7 | 18-6 | 17-4 | 16-3 | 13-7 | 8-3 | 3-0 |
| 3 | 8-5 | 20-6 | 22-4 | 24-2 | 24-2 | 22-7 | 22-2 | 21-2 | 19-1 | 13-2 | 4-5 |
| 4 | 9-5 | 22-2 | 23-4 | 24-4 | 24-4 | 25-0 | 24-5 | 24-0 | 22-3 | 18-2 | 5-6 |
| 5 | 10-2 | 22-6 | 23-6 | 24-6 | 24-6 | 25-4 | 25-1 | 24-4 | 23-3 | 19-5 | 6-0 |
| 6 | 10-0 | 21-7 | 23-2 | 24-0 | 24-2 | 24-4 | 24-1 | 23-3 | 22-1 | 16-7 | 5-1 |
| 7 | 9-2 | 19-5 | 21-2 | 22-4 | 23-0 | 22-4 | 21-7 | 20-4 | 16-4 | 7-4 | 3-2 |
| 8 | 8-1 | 16-2 | 18-2 | 19-2 | 2-0 | 19-7 | 18-1 | 17-5 | 0-6 | — | 0-6 |
| TRAN. | 7-2 | 15-1 | 17-4 | 19-1 | 20-0 | 18-3 | 16-6 | 11-2 | 0-6 | — | 0-6 |

13' 10" x 4' 2" DISP. 472# P.C. .55

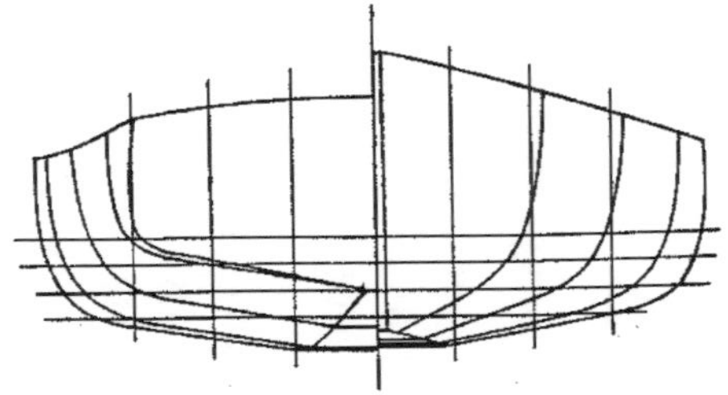

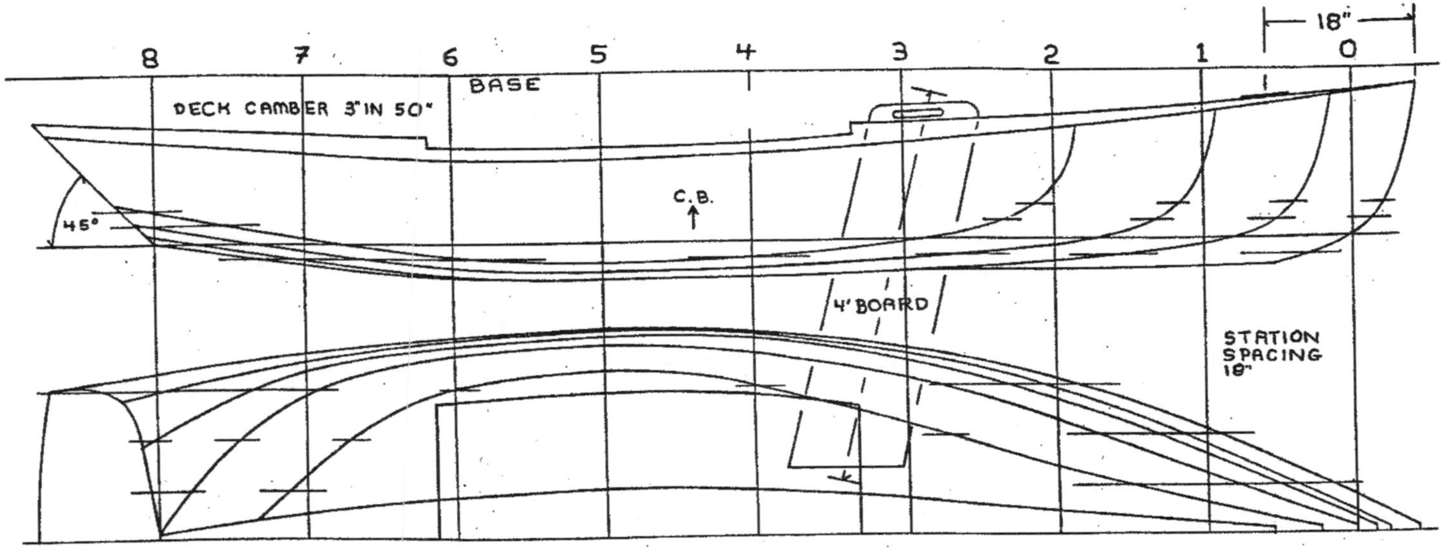

Pleasure Melonseed lines and offsets

very elaborate mold to make when only one hull is to be pulled from it; and in spite of his long experience he says that he still has agonizing problems with the planking sticking to the stringers, making it difficult to lift the hull off the mold.

I decided to use sectional molds only—no stringers—but wasn't sure how far apart the molds could be without the planking forming flat spots between them. A couple of years ago a friend built an Ian Oughtred dinghy, 7 $^1/_2$ feet long and sharp-bowed, and he used only three molds, stem, and transom, so his molds were 22 $^1/_2$ inches apart. But the tubbiness of the little boat bent the plywood quite severely and guaranteed fairness. With the relatively slight bends of the Melonseed amidships, I spaced my molds only 14 inches apart. The hull did come out fair, but the close-spaced molds made it awkward to get inside after each plank was hung to scrape out the excess glue.

For a bottom board I used 9-millimeter sapele plywood and for the clinker planks, 5-millimeter. Okoume was the other possibility and would have been lighter, but durability counted for more with John. Garaged, I'm sure that an okoume boat would last many decades, but in the end it is proper that the owner make these decisions. As is, the complete but unpainted sapele hull and deck weigh 128 pounds. A Melonseed takes less than six sheets of ply to build but several hundred hours of time, so it sure would be a shame to use fir or lauan marine ply, let alone A/C ply or underlayment.

I scarfed the planks on the boat. They could have been scarfed on the bench, whole sheets at a time but would have cut to more waste that way; and planks 14 feet long and 5 millimeters thick, covered with epoxy glue, need at least two people to handle them. I clamped the laps with C-shaped plywood clamps cut from $^1/_2$-inch plywood (here's a good use for lower-grade ply) and driven wedges. They do not have the holding power of screwed clamps, but they have enough for this work and can have any depth of throat that you like. Building the hull upside down, gravity makes them hold less well on the topsides than on the bottom. At the turn of the bilge, the clamps must grip a beveled surface. Their holding power can be improved by wrapping each wedge with a scrap of fine sandpaper. After each use, the gripping surfaces of the clamps should be inspected, because if any glue has gotten on them they grip less well next time.

The scarfs I clamped with screw clamps, over ply butt blocks to spread the load, over waxed paper as a parting agent. On the molds I used a dab of cellophane tape under each lap, and Carol and I lifted the hull off the molds with no trouble. Susanne Altenburger suggests that the laps could be clamped with temporary drywall screws, and of course they could. That would leave holes to be filled and sanded later, but there's plenty of filling and sanding on a boat like this anyway. However, the pressure needed to get each screw started—especially in hard sapele—can easily close up the lap, making it perhaps $^5/_8$ inches instead of $^1/_2$ inch, which distorts the hull noticeably.

For all kinds of clinker building, I prefer a dory lap at the ends of the planks to a rebated lap. Certainly it is easier to make, and I suspect that it is stronger, although that could probably be argued. Perhaps a rebated lap has a better chance of keeping water out in a boat with no epoxy or stickum between the planks. But how many of us are building that way these days?

Clinker plywood planking with plywood clamps;
screw clamps join a scarf

Unavoidable holes for filling later were in the bottom, for example, where I screwed the bottom plank to the molds to keep them aligned while the clinker planking went on. They won't shorten the boat's life, but are messy work to fill and dusty work to sand, especially as the epoxy putty must stand proud of the surface initially, because it shrinks while setting up. For years Dave Carnell and I have been discussing the best filler to make a putty of epoxy resin, both for gluing and for sealing holes and screw heads. It seems that nothing could be worse than fumed silica, which is hard as steel for sanding but so light weight that while being mixed with resin it floats up into the builder's lungs. And yet to date nothing seems better.

Wood flour is often not fine enough, and in a vertical join the resin dribbles out of the flour, leaving a starved join. Friends tell me that cotton fibers do the same. Dave once touted pulverized limestone, which doesn't float in air and doesn't slump in the join. But it is stone, and it takes the sand off sandpaper at a single swipe. Recently he sent me a sample of tilesetter's talc. It doesn't float or slump, and it sets no harder than silica, but it's tricky to mix with the resin: too thick and too thin are just a few grains apart. Years ago I had success with Johnson & Johnson baby powder, but the sanding smell is nauseating. Wheat flour also works well, but I mistrust its longevity. So I'm still using silica in the joins and the easier-sanding microballoons for filling holes and fastening heads.

Building the glued-lap Melonseed, the ply cut very economically. Each spiled plank seemed to nestle in the hollow of the plywood sheet left by the last, and it took less than three sheets to plank the hull. A shape that large in chined plywood probably would have used more sheets. However, the work went slowly. On the best day I hung only three 7-foot planks, and the theoretical maximum on a 14-foot hull is four planks, because the clamps get in the way of hanging another one. Most days I hung fewer. No matter how carefully the glue is scraped from the joins on the outside, there is still some sanding to do the next day; and when the hull was turned there were 12 hours of very unpleasant chipping and sanding to remove the gobs of excess glue that had hidden themselves under the clamps and molds. Come on, Dave, find me a better filler soon!

I decided to put half-frames in the boat. They may not be necessary, but they don't add a pound of weight to the whole structure, and they are insurance. Coming down only around the turn of the bilge, they are wholly underneath the side decks and aren't in John's way. They seemed a good idea because, in the bottom and the topsides, the laps have a $^1/_2$-inch glue join. At the turn of the bilge where the planks meet each other at sharper angles, the glue join is less, and much of it is end grain ply. An epoxy fillet would have strengthened those joins, but fillets are not so easy to make tidy, and they create more sanding and smudge the decorative clinker look. Carol and I steamed the half-frames, securing them with screws driven through the planks. Several days later when they were thoroughly dry and set they were removed and epoxied back in place. They must have complicated the painting, but that was John's part of the work.

According to an article in the *New Yorker* magazine many years ago, glued-lap clinker plywood is a method thought up by a house architect who built a good many boats and had no trouble finding customers for them. But then he grew tired of it and passed the technique and his molds on to others. Frankly, when I started the Guidera boat I had no idea how many hours it would take and how many of them would be spent sanding. The method may not be as time-consuming as cold molding (which Susanne Altenburger calls "gold molding"), but it's plenty tedious enough. Some of Tom Hill's designs are V-bottomed with clinker topsides, and that does save some labor; but the resultant hull looks like a clinker boat only from some angles and goes through the water no better than other V-bottomed hulls. The method does make a very light, strong, stiff, attractive hull. For an amateur builder who doesn't care how much time he spends, it has much to recommend it.

Melonseeds have highly crowned decks: 3 inches in 50 inches on this one, compared to about 1$^1/_2$ inches in 72 inches on *Mayfly*. The original ones had sawn deck beams, but laminated beams can be lighter and give better access to the interior. Any laminated member springs back to some extent when the clamps are removed, but Joe Dobler came up with a good formula for predicting it. If $N$ is the number of layers, then springback equals 1 divided by $N^2 - 1$. The Melonseed beams use five layers of stock, so the springback should be 1 divided by 24, or $^1/_8$ inch in the 3 inches of crown. Dobler's prediction was spot-on.

For the spars I chanced on a local supply of Sitka spruce at least thirty years old. Although *Skene's* says it's the best spar stock, it is less strong than Douglas fir and spar dimensions must be increased 10 percent, so little weight is saved. It costs

John Guidera's Melonseed (Photo courtesy John Guidera)

three times what fir does. This batch of it was straight-grained and pleasant to work, but probably thirty-year-old fir would have been pleasant, too. The spars are a lighter color than fir spars would be, and because the wood is softer they nick more easily and hold fastenings less well. Like fir and many other softwoods, Sitka can spall if flat-grained boards are used.

John made a nice job of finishing his Melonseed in white and tan, with varnished spars and sheerstrake. He brought it over for trials, and we were both pleased with the way it sailed. The foot-deep rudder turned it promptly and reliably, though the 5 $^{1}/_{2}$-foot tiller is, of course, not very responsive and has to be swung a good distance, and I would have preferred an untraditional hiking stick. John later tried sailing his boat against two Bluejays, which are about the same length but with a larger more modern rig and cruder hull form. He licked one of them. He had no trouble sailing circles around a Beetle Cat. He says the boat is less lively with two aboard, and I did originally draw it with a rig about 10 percent larger and a foot higher than the plans show. Probably he was right to insist on the smaller rig: the boat is relatively narrow for her length, and the wide side decks limit the skipper's ability to shift his weight unless he hikes out, which certainly isn't traditional.

Kneeling and facing forward seems to be the best way to row the Melonseed, and probably that's how the working ones were rowed, although a neighbor of ours who built one with Chapelle's low freeboard simply uses a kayak paddle. Probably John would find his rowing more comfortable if he carried a little stool to sit on, but that would clutter the boat, and he does sail her 99 percent of the time.

Sail- or motor-powered boats should have some auxiliary power, and muscle power is the most reliable. The Dobler rows very well for a sailboat, and Al Whitehead carries a canoe paddle in *Mayfly.* The mini-garvey has nothing but her daggerboard to get us ashore, but the river is narrow and the tide is always moving one way or the other. Even *Puxe,* our 22-foot motor launch, carries a couple of canoe paddles in her forepeak, and we do use them several times a season, although often just to back out of a creek that has suddenly become too winding and shallow for the outboard.

Like the Bobcat, John's Melonseed is a great satisfaction to me because it is enjoyed and used often. Probably both these boats will be used more hours than I spent building them, and that's a fair yardstick to set against any boat. Both of them excite admiring comments whenever they are launched, and that's a satisfaction to the owners. It puts me in mind of what I said at the beginning of this chapter, that a daysailer can give more pleasure than any other kind of boat.

3

Runabouts

# Runabouts

Twenty years ago Robert Speltz published the first of a long series of books called *The Real Runabouts.* They are big glossy books full of color pictures of big glossy boats. Looking through them, it's easy to conclude that *real* and *expensive* are synonyms. Compared to the Speltz boats, the four in this chapter could easily be called unreal runabouts, but that does not seem to interfere with the pleasure their owners find in them.

Are there more canoes or more runabouts in America? It's hard to say because canoes aren't registered in most states and people keep them forever, just on the chance they'll be wanted again some day. But even Dave German, who lives across the river from us, finally decided he'd had enough of his fiberglass runabout with a soft transom and seized outdrive and advertised her for sale. "Find a buyer yet?" I asked him.

"Think so. A guy from down county says he's going to give me twenty-five dollars."

"Don't be a chump. Call him up and tell him you'll give him twenty-five dollars to come take it away." But Dave didn't, and the guy didn't come, and several years later the boat is still sitting at his place out on the highway with a sale sign on it. The honeysuckle seems to like it, and in a few more years no one will remember it.

A runabout is an open motorboat designed for short trips in sheltered water. Today we think of them as planing: lifting their whole weight and bulk out of the water and skimming along on top. By conventional definition that happens at three times the square root of the waterline, but most boats that go that speed or even faster aren't really out of the water; you can tell because they make a big wake. Scott Stetzer, an Atlantic City newspaper photographer, was for some years the sparkplug of a garvey club that met every week of the year for at least a short excursion in their boats, and they liked to come visit my shop whenever I called Scott and told him I had a garvey under construction.

Once I went downriver with them in our 22-foot *Puxe,* past Yank's where the solid land falls away and there is an immense marsh laced with rivers and creeks. We were making 10 knots, *Puxe*'s cruising speed, and the garveys were disturbing the water more than I was. They were fiberglass 16- to 18-footers with rope-start outboards up to 65 horsepower, no batteries, and no remote engine controls except a piece of plastic pipe attached to the tiller with hose clamps. The skipper stood up amidships, the pipe in one hand and a rope from the bow in the other.

Below Yank's they turned on the throttles, and the boats literally leapt out of the water like startled ducks from a creek. They were probably doing 50 miles an hour, and the acceleration was almost instantaneous. The wakes literally disappeared. From where I sat, putting along with my mouth hanging open, it seemed that all that disturbed the water surface were the legs of the outboards. To do that, what's needed is an absolutely flat bottom aft and 1 horsepower for every 15 or 20 pounds of load. At that time the garvey club was planning a summer cruise to Canada: up the Jersey coast and the Hudson, across the Erie Canal and Lake Ontario, a round-trip voyage of a thousand miles. They expected to observe the speed limits where there were any, see the sights, camp out in their boats, and be home again in a week.

A boat should plane with 1 horsepower for every 40 pounds of load, and a perfectly flat bottom can do as well for 50 pounds, but other factors come into it. Motoring across Florida's Biscayne Bay in my brother-in-law's grossly overloaded 60-footer, I taunted him about how his hull was supposed to plane. He kept his peace, but when we reached a cut where the water was only a couple of feet below the keel, he backed off the throttles until the boat's stern wave came up under the hull, then jammed the throttles open. "She's up!" he yelled down to me triumphantly. I looked over the side and saw that she was in a sense up, going about a knot faster than she had been. She was throwing even more wake.

Early motorboats did not aspire to plane. Few steam engines produced as much as 1 horsepower for each 50 pounds of engine and boiler weight, let alone the weight of fuel, boat, and crew. Early internal combustion engines weren't very much better, so the few boats that exceeded hull speed cut through the water with their long, narrow shapes. What changed all that was the price of fuel and the power-to-weight ratio of engines. Fuel is far cheaper now in real dollars than it was a century ago, and we have fought at least one war to make sure it stays that way. Today a 50-horsepower outboard weighs less than 200 pounds, and inboards don't weigh that much more, even the diesels. So why not just skim across the water?

It may, therefore, seem odd that of the four boats in this chapter, only the Ocean Skiff planes. *Minnow* did plane when she was new, and *Speed King* will plane someday, if a 10-horsepower engine replaces the current 3. So only Emil's flatiron has no planing in her past, present, or future, and the why of that is an interesting story.

# Emil's Flatiron

Emil Williams grew up in a boating family. His uncle introduced him early to the creeks and swamps of central New Jersey and to the art of bobbing for eels. They dug earthworms, a great many of them, and laboriously threaded them longitudinally until they had strings of worms long enough to roll up into balls the size of baseballs. These were called *bobbins.* At night they stole out in a cedar flatiron and dangled the bobbins in fresh water.

When they were very quiet and still (and also lucky), eels would take the bobbins, swallowing them whole. The eels were pulled aboard and the undamaged bobbins were extricated from their throats and sent down to bob for other eels. The eels were taken home and pickled, to be used later for trap bait; but as everyone enjoyed the bobbing more than the trapping, the pickled eels tended to accumulate. One of Emil's early memories is of his uncle's garage with shelves full of eels in Mason jars. Some had been there long enough to have lost their color.

As an adult Emil found other reasons to take him to the water. He studied nature photography and became good enough to have his pictures published in *New Jersey Outdoors.* His vessel continued to be a cross-planked cedar flatiron, no doubt hammered together with hardware-store nails from a design in an ancient and yellowed *Popular Mechanics.* When this vessel swole up for the last time, he came to see me.

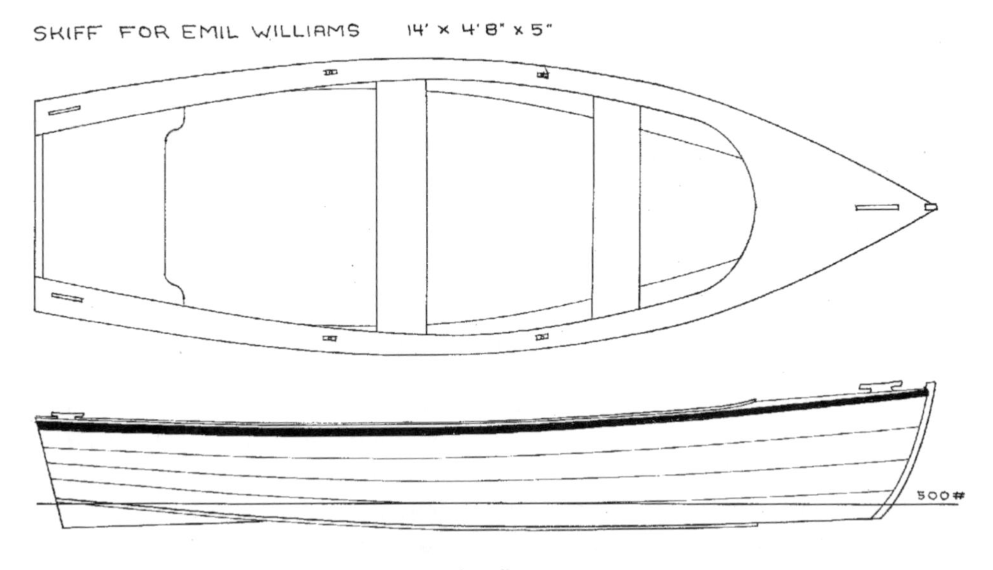


Emil's flatiron

He wanted another 14-foot flatiron, but wider and with more freeboard and tight from the moment it rolled off the trailer. He wanted a short foredeck for dry stowage and prettier looks and a better finish than the old boat. He wanted to be able to pole it, row it, or power it with the 4-horsepower outboard that he already owned.

A flatiron is any flat-bottomed boat, pointed at one end. Designing one is not a big chore, especially if it is to plane. The bottom then can be perfectly flat fore-and-aft as well as athwartships. Add more people and the boat just sinks deeper in the water until power is applied, which brings it up on top of the water. On a plane there's little to choose between it and a garvey, and naturally the flatiron is easier to build. No doubt there is an ideal transom width, which could be related to the cube root of displacement or of horsepower; but that can be estimated pretty closely by looking at a few other designs. It's amazing how much discussion goes on about these simple boats and how many different but similar designs are on the market.

Designing a displacement flatiron is only a little more complicated. Weight must be estimated and the bottom rockered to bring the transom clear when the boat is loaded and at rest. For best performance both ends should be finer than for a planing hull, which also makes a prettier boat. The smaller the power, the narrower the transom should be, and as already discussed in the first chapter, if the power is as feeble as human muscles, the transom should disappear altogether, at least at waterline. Most displacement flatirons have some twist to the sections to give them a narrow entry and reserve buoyancy, and indeed Emil's does too, although I'm not sure how much it matters. It's really just a scale-up of Robert M. Stewart's *Susan* skiff, rockered to suit the weight estimate and with the extra freeboard. With 4 horsepower and 500 pounds to push, Emil's boat has far more power than a freighter but not nearly enough to exceed hull speed.

We decided on a plywood bottom because it would stay tight. The clinker cedar topsides are for looks, as is the round forward coaming and the widening side decks aft. With thwarts glued in place the side decks wouldn't have been necessary, but Emil wanted the thwarts to lift out. The two stringers on the bottom add wetted surface for rowing or poling but keep the sole uncluttered for walking. For the poling Emil wanted a good, big skeg.

Like all my designs, this boat has a false nose. A rebated stem was the ticket in the days before waterproof glue, but why are people still doing it? It's laborious to cut, wasteful of lumber, and tricky with the bevel changes of a round bottom, although not with a flatiron. Planking must be spiled to a rebated stem, but can be run wild and cut back later with a false nose. Keel rebates are also better glued up than cut in.

The boat went together without trouble. Although I don't usually keep a time log, I disciplined myself to do so this time, to see how the time spent compared with the estimate that was used for the bid price. It was 5 percent over, and it would be nice if all jobs came out that close. Even when they don't, the owner of a small shop is not losing money; he's just earning less for each hour worked, and the work is still enjoyable, so there really is no loss at all. Only a bad estimate of materials can get a small builder into real trouble, and materials are easier to estimate than time.

Emil Williams in his flatiron

Emil made a wonderful job painting his flatiron. He probably put more hours into the painting than I did into the building, because he brought at least one other color scheme nearly to completion before deciding that he didn't like it. The final scheme is medium blue thwarts, medium gray deck and guards, and lighter gray topsides and interior. The bottom is red—enamel, not anti-fouling—and the red is brought up the stem post and all the way aft in a one-inch band below the guards. The gray gives a workboat feel, but the detailing belies it.

In his enthusiasm for his new boat and wanting the best for her, before launching day Emil traded in his old 4-horsepower outboard on a new 10. That must have cost nearly as much as the boat, and as the picture shows, it doesn't suit the hull at all. Not only does it double the weight hung on the transom, but its extra power pushes the stern down deep into the water with negligible increase in speed. If this boat had been designed for 10 horsepower, with a wider transom and no rocker, it would certainly have planed out flat with one person aboard, although it would have rowed and poled less well. With 10 horsepower and rockered bottom, she could be throttled back to give 4-horsepower performance, but it's against all our natures to do that. I tried, but it wasn't long before I was twisting the throttle a little harder and a little harder, shortening up the waterline at the bow and digging a hole at the stern, bemused by the noise and wake and the illusion of speed that they gave. Maybe Emil has better self-control.

Emil's had his flatiron for ten years now, and he may have given her another color scheme because he seems to like doing that. For some time he kept saying he would bring the boat down again and take us for a longer ride, but I think he knows that out in the marsh the distances are long for a boat with less than 5 knots top speed; and I think he's embarrassed about the engine. The boat is still a fine platform for taking pictures of nature, and I hope he gets out in her from time to time. I'm sure he keeps her in apple-pie order.

# Minnow

*Minnow* was a Keystone runabout 13 feet by 4 $^1/_2$ feet, built in Chester, Pennsylvania, in 1939. She was a fully planing hull with no rocker and a hard turn of bilge. Her straight sheer probably simplified construction in some way, and I suspect that she was built over a jig and very quickly in a shop with four or five workers. Tumblehome aft was her one concession to esthetics.

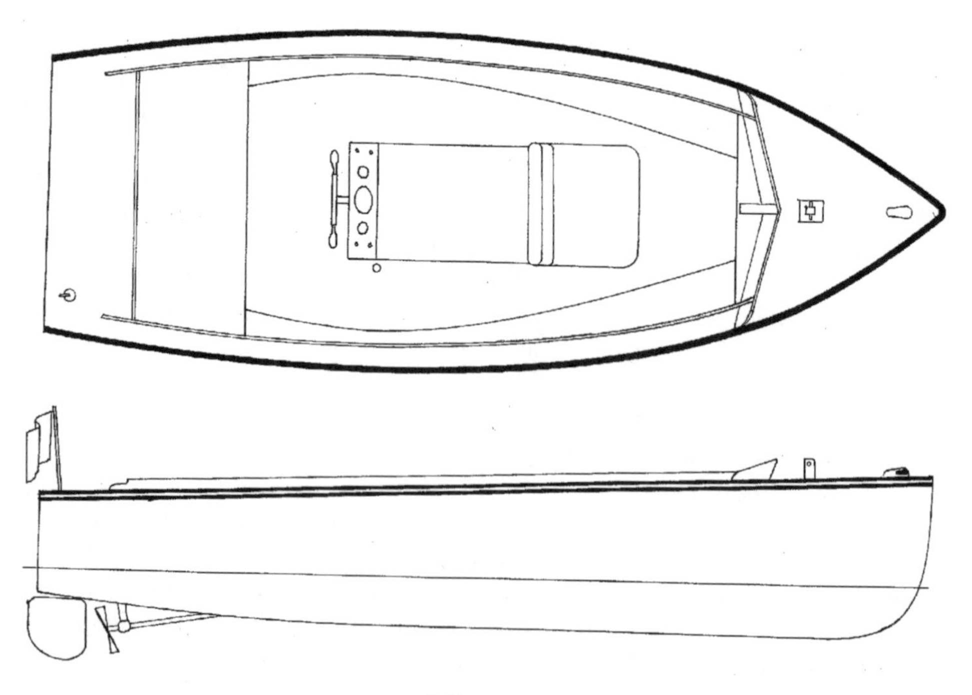

*Minnow*

She had a 1-inch transom to take the thrust of an outboard, but the rest of her was very light: steam-bent oak ribs $^3/_8$ inch by 1 inch on 6-inch centers, $^3/_8$ -inch clinker cedar planking, copper rivet fastenings. She steered with cables from a bench seat forward. She certainly was light enough to plane, even with the heavy outboards of the period. In 1939 America was still struggling out of the Depression, but a few working men were beginning to have money, and perhaps *Minnow* and her sisterships were aimed at them.

Her original owner brought her to the Jersey shore. He must have liked her because he kept her twenty-five years, but he soon decided that he preferred inboard power. Perhaps he tired of the finicky outboards of that era, or perhaps he came upon a 10-horsepower Kermath for a good price at a time when he needed an engine. For whatever reason, that's what he put in her.

Kermath made marine engines in Absecon, New Jersey, but so long ago that no one in that town now remembers where the factory was. *Minnow*'s engine is a side-valve twin, water-cooled, with side-draft carburetor and magneto ignition. The flywheel is on the front of the engine and its teeth and starter are not enclosed by metal, but only by wooden furniture. Nonetheless, little gasoline engines like this one are pleasanter company than the diesels that have replaced them. They are smoother and quieter, due to their low compression. They burn a few more ounces of fuel per hour, but they are so much cheaper to manufacture that it would take many thousands of hours of running time to make up the difference at the fuel pump. In an open boat a gasoline engine does not pose the explosion threat that it does in a closed one.

There is no reason to think that the Kermath was new when it was installed in *Minnow,* and some of the $^7/_8$ -inch running gear may have come with it. The installer set out to make a permanent job of it: into the bilge he fitted two four-by-fours, the width of the engine mounts apart and almost as long as the boat. They were attached to the bent frames with myriad tiny sheet-metal straps and screws. They were then covered with sheet steel, and the sloping engine beds were built on top of them. There is no possibility that after the installation of the Kermath *Minnow* ever planed again, and that's the reason she lasted. Lightly built clinker planing hulls have a very short life because they flex. Not much water comes in through the seams, but eventually some sand and dirt come in with it, and clinker seams are difficult to clean out and caulk. The sand and dirt hold the seams open, allowing more water, sand, and dirt to come in. The harder the boat is pushed in rough water, the sooner it happens.

Even *Minnow* with her Kermath leaked eventually, and various steps were taken to deal with it. On the inside the planking was packed up flush with the ribs, bow to stern and keel to waterline. This must have been a tedious job, cutting little blocks and installing them with tiny screws. Both wood and Masonite were used. On the outside the entire bottom was glassed, which is an extraordinarily tedious job on a clinker hull, and a surprisingly good job was made of it. None of this completely stopped the leaks, and all of it drove the boat farther down into the water. *Minnow* was now dragging several square feet of transom, which made her awkward to manage as well as slow, and a squat-board was bolted to the transom.

If it had been shaped to be an extension of the bottom of the boat it might have helped, but being only a flat plank it was 4 or 5 inches above the bottom at centerline, and I doubt it accomplished much except to be hideous.

When *Minnow*'s owner died the boat lay around a marina for several years, was bought by a man who used her one season and then returned her, and lay around several years more. By 1970 there wasn't much interest in 5-knot runabouts, and the marina was thinking of cutting her up when she had the good luck to be spotted by Sam Evans. Sam was thinking about a small boat to poke around the lagoons behind Avalon, New Jersey, on summer evenings, and he told the marina to fix her up first class.

Denny Bock was the mechanic who took on the job, and he at once fell in with the spirit of it. Improvisation was the key because Universal had bought the Kermath company but didn't have all the parts for the engine, let alone a shop manual. Nevertheless Denny did succeed in rebuilding it, and he painted the block with blue engine enamel and picked out the cast Kermath lettering in red. He installed a wheel steerer with quick 1:1 gear and a shaft to connect it to the rudder. He converted the electric from 6 to 12 volts and fitted as many gauges and switches as he could think of. He installed a chrome-over-bronze shift lever so big that it towered over the console. Eyeballing it, he decided to cut it off 6 inches above the base and bolt it together shortened, thus preserving both handle and shaft fitting. Looking over his handiwork twenty years later, he smiled at me ruefully. "Sometimes when you do these things you forget how long they're going to be around," he said.

A house carpenter installed a teak deck. He made it a work of art, as many carpenters will do if given enough leash, especially in wintertime when work is slow. A friend of ours who wanted a shed where he could practice his hobby of building small traditional boats was persuaded by the shed carpenter that the details and finish should match those of his Victorian house and that for looks the shed should be square. So now he has a shed best suited to building multihulls and not long enough for what he wants to do with it.

*Minnow*'s deck carpenter started with a layer of $^1/_2$ -inch A/C fir plywood; over that he laid $^1/_2$ -inch teak strips, without bedding and fastened with paneling nails. He gave her a big teak foredeck coaming and a coaming knee so huge that its dimensions can only have been based on the size of the hunk of teak at hand. On the aft deck he may have run out of paneling nails, because the fastenings there were salty-looking oval-headed sheet-metal screws. The new deck may have weighed as much as the original hull, but Sam was not deterred. For him the preposterousness of *Minnow* was part of her charm.

He enjoyed *Minnow* for many seasons. He kept her in a slip by his house, and every winter Denny winterized the Kermath and hauled and stowed the boat indoors. Despite this care, the glass and Masonite were not improving the cedar planking, and the four-by-fours sweltered under their sheet metal. One winter the lifejackets were left behind in the forepeak, and mice spread kapok throughout the bilges. The boat continued to leak some, and the kapok was very troublesome in the bilgepump. Somehow Sam got to thinking that the kapok was alpaca. "Everything will be all right" he used to say, "if we can just keep the alpaca out of the pump." Nevertheless he and his new wife had many happy evenings in the boat, putting around the lagoons and looking in windows to see what other people were doing. Janet was a big woman who liked to laugh, and I think *Minnow* played a part in their courtship.

But the boat was rotting, as fifty-year-old boats do, especially when glassed over and slathered with packing and decks and furniture not properly caulked or bedded. At spring launching in 1990 the water came in too fast for the pump. Denny lifted the sloping wooden cover above the stuffing box, and the keelson was the consistency of oatmeal.

The marina where Denny now worked and *Minnow* now wintered made its living selling and servicing fiberglass fishing boats, many of which were kept in racks and launched with a fork lift. Few employees knew about wooden boats or cared about them, so I was called down. A bunch of us stood around glumly, elbows on the teak side decks, staring down into *Minnow*'s rotten innards. I told them that although I could build boats, I knew very little about repairing them and so could only suggest building a new hull. To make matters worse, I had work lined up that would keep me from starting before November. Sam didn't say so then, but Janet had breast cancer, and of course she was having every treatment and was expected to recover, but he very much wanted to have the boat that season.

Sam thought it over for several weeks and may have consulted boat repairers whose phone numbers I gave him. He knew he could buy a new 13-foot runabout, engine and all, for less than a new hull for *Minnow* would cost. But it would be a boat without character, without any associations to past pleasures. There wouldn't

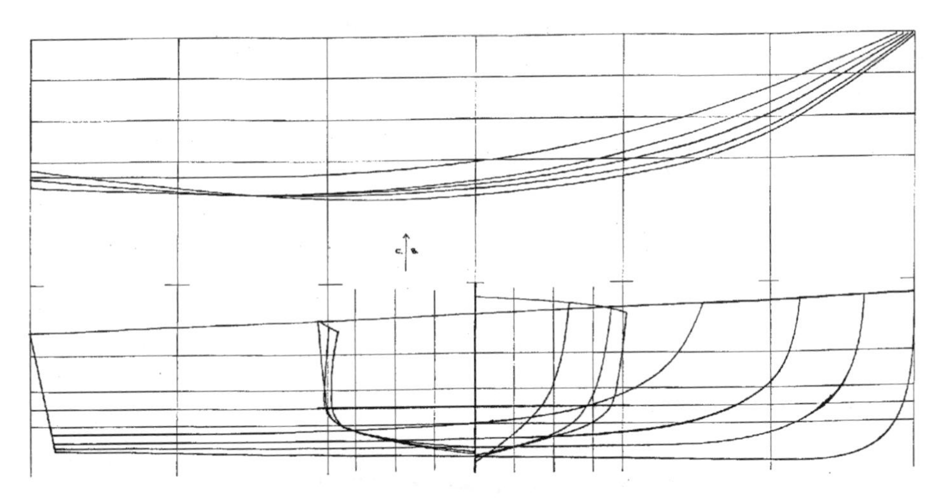

*Minnow's* original lines

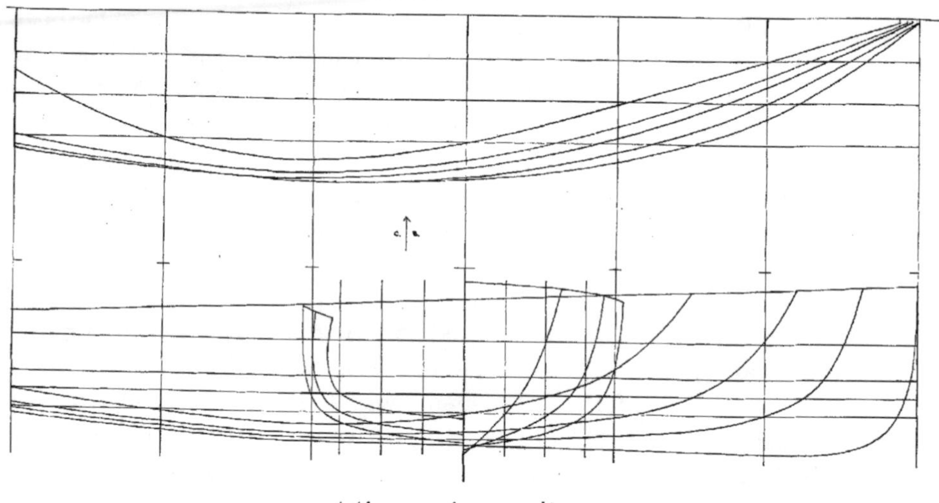

*Minnow's* new lines

be a seat for him aft behind the wheel, looking down at the gauges and switches. There wouldn't be Janet's seat forward with its padded backrest. There wouldn't be teak or the intricate pattern of clinker planks and steamed ribs. In the end he decided on the "sentimental gesture" as he called it, and we came to terms.

When I brought the hulk up to my shop the next winter, my first job was to take off the lines. As far as possible Sam wanted the look and feel of his boat preserved. He wanted the materials and finishes preserved but was willing to let me change the underwater shape if I was willing to take the trouble. He agreed that the new hull could have a plumb transom instead of the raked one that the outboard had required. That gave a precious 4 inches more waterline aft, which wouldn't change top speed significantly but would allow me to work in more buoyancy to float the immense weight that the boat had accumulated over the years. From the start I was determined to float the hull high enough to eliminate that gruesome squatboard.

I took the lines by marking sections on the hull with a crayon, spiling them onto a curved spiling board and from there onto the lofting panels. Drawing the lines from the sections and fairing them up was a kind of lofting in reverse and interesting to do. It turned out that the hull had hogged $^3/_4$ inches but was otherwise pretty fair and symmetrical. In 1990 her waterline gave a displacement of 1,175 pounds. In 1939 the hull probably weighed 150 pounds and the 10-horsepower outboard 100 pounds There was no getting back to that, but I could save something in the glassing, the packing, the multiple layers of $^1/_2$ -inch plywood floorboards. Sam wanted the old teak deck saved if possible; but if not, a new thinner one would certainly weigh less and would lower the center of gravity agreeably.

I started with the same midships section but rockered the lines up aft, reducing the area of immersed transom by about 70 percent. To keep the center of buoyancy where it had been, the extra 4 inches of waterline were a help, and I increased the deadrise forward and slimmed the waterlines—but not too much, because the sole still had to be wide enough for walking around the console. The new lines reduced displacement only 53 pounds, but still allowed the full waterline length if she floated a couple of inches light of her marks.

I set up molds and notched new four-by-fours into them. The engine didn't need such massive supports, but they gave a landing for the floorboards and a reference point for all the furniture and the electrical and mechanical placements. In western red cedar they weren't terribly heavy. Usually I prefer eastern white cedar because it is less splintery and can be bought from local sawyers, which is fun. But it is green and thus needs to season a year for each inch of thickness, so kiln-dried western red seemed the better choice. I ribbanded out the molds and bent the frames around. When the frames were dry I glued and screwed them to the four-by-fours and to the keelson.

Clinker planking with boards is both easier and more fun than with plywood. The boards can be full length so there's no scarfing, and if you buy flitches the curve of the plank often conforms to the curve of the heartwood in the flitch. For *Minnow* I used stock just over an inch thick that I had bought from a sawyer as soon as Sam and I reached agreement and put up in the shed rafters to season with stickers between the boards. I spiled out each plank on a 14-foot spiling batten and after it was cut to shape resawed it in the bandsaw to get a plank for each side. My 12-inch bandsaw with 7-inch throat height and maximum $^1/_2$ -inch blade width really isn't ideal for resawing; but with a big square block of wood clamped to the table for a fence and a sturdy comb to hold the plank against the block, it can be done.

At that time I had no thickness planer to finish *Minnow*'s planks, but like all power tools that one doesn't have one learns to substitute other tools: a maul plane, a sander. The work goes more slowly but it does go. Finally you buy the missing tool and wonder how you lived without it. I did build a number of boats before buying a bandsaw, but now I can't imagine being without one.

When a hull is upside down it's easy to get in trouble by labeling parts for it "port" or "starboard." When it's time to attach them you may have a different take on what's port and what's starboard, and that may cost a good deal in time or materials or both. It's better to think of sides of the hull as north or south, east or west, depending on the orientation of your shop, and avoid the nautical terminology until the hull is right side up.

As with clinker plywood work, screws are the best fastenings for cedar laps if the stock is thick enough to hold them. That's especially true if you're working alone. In fact, the little experience I've had with rivets and roves suggests that they hold a join but are not much good at pulling it together. Yes, the shaft of a rivet is supposed to compress with peening and pull the join, but it's a feeble pull compared to the power of a screw thread. Rivets and roves have an antique look because they were used before the technology existed to cut screw threads by ma-

*Minnow's* new hull

chine. But probably in Noah's ark, as in the Jesus boat from the Sea of Galilee, the planks were held together by tenons every few inches along their edges. How antique need we become?

Polyurethane stickum was used in every inch of every seam of *Minnow*'s new hull. I am no longer sure that it remains flexible forever, but for a very long time it does. It also retards the entry of sand. In his legendary canoes Henry Rushton shellacked the seams, that being the best material available to him, but I doubt that the original *Minnow* hull had anything at all in hers.

The spacing, or lining off, of clinker planks is often discussed, so probably there is no best method. Of course, my bandsaw's 7-inch throat height limited what I could do, the more so as the topmost plank should be wider than the next one down by the amount that the guard rail overlaps it. Probably it is good to have planks at the turn of the bilge narrower than those on the bottom or topsides, and the harder the turn, the narrower they should be. Topsides planks should be the same width amidships, and all planks must have tapers at the ends that echo each other, or a clinker boat will look a mess even to an unpracticed eye.

While the new hull was being built, Sam and Janet came down often to see the progress. He would look wistfully at his old *Minnow* and try to be cheerful about the new one. Often Janet, who was having chemotherapy, would sit up in the livingroom with Carol. Once she said, "I never was sick a day in my life. This cancer is a pain in the ass!"

It turned out that the old deck could not be saved, although the coamings and the knee could. It would only come off the old hull in pieces, and the A-C plywood was in sad shape. I did sneak a little camber into the new one, using $^3/_8$ -inch marine ply and gluing down to it teak strips $^1/_8$ inch by 2 $^1/_4$ inches with $^3/_{16}$ -inch seams between the strips. Teak glues better if it is first wiped down with acetone to get some of its oil off the gluing surface. Until the epoxy was nearly dry the strips were held in place by steel staples shot down over heavy cotton twine, which made the staples easier to extract. The seams were payed with black polysulfide.

The trickiest boatbuilding that I ever did was putting the old *Minnow* fittings into the new *Minnow* hull. Building a new console would have been quicker than installing the old one, but there had to be *something* of the boat that genuinely was old. The console had to go down over the engine within $^1/_8$ inch of where it had been or the throttle linkage wouldn't hook up, and I couldn't put a cable throttle in because it wouldn't have felt like the old rod throttle. The steering linkage also had no adjustment, so the back of the console had to be *exactly* the same distance from the rudder post. Many a trip with ruler in hand was made from the old hull in one end of the shop to the new one in the other and to the lofting boards and back again. One change and improvement that Sam did want was a new steering wheel, because what she had was one of those lauan wall ornaments the color of well-masticated bubble gum. The smallest bronze wheel from Lunenburg Foundry fitted her nicely.

Meanwhile the Evanses came less often, and on the phone Sam sounded sad and tense. "Janet's a very sick girl," he said. But *Minnow* was looking more and more her old self. The day came when I could pull that loathsome old hull apart—it needed very little cutting—and cart it to the dump. Sam once asked what I had done with it, but I answered evasively. By spring the boat was ready to trailer down to the marina, and Denny greeted her with jubilation, particularly admiring the absence of the squatboard. He tuned her up, and I went down the day Sam was to motor her to her slip in Avalon. But he had already left, hurrying to get home, and that was the only time I ever saw her under way, in the distance carrying her old captain down the inland waterway.

Sam was delighted to be done with the alpaca problem, but despite the effort I made to duplicate *Minnow*'s looks and feel, he had trouble thinking of the new hull as the same boat, and he still tends to talk about *Minnow* and "the new one." He thought it might be a knot faster, and it made less wake and steered easier. He was happy that Janet had four or five rides in her, although I suspect that each of them was going out to show the other that they were enjoying it. When Janet died Sam's interest in *Minnow* diminished, but he did keep her in the water several more years, and I hope he got some pleasure from her.

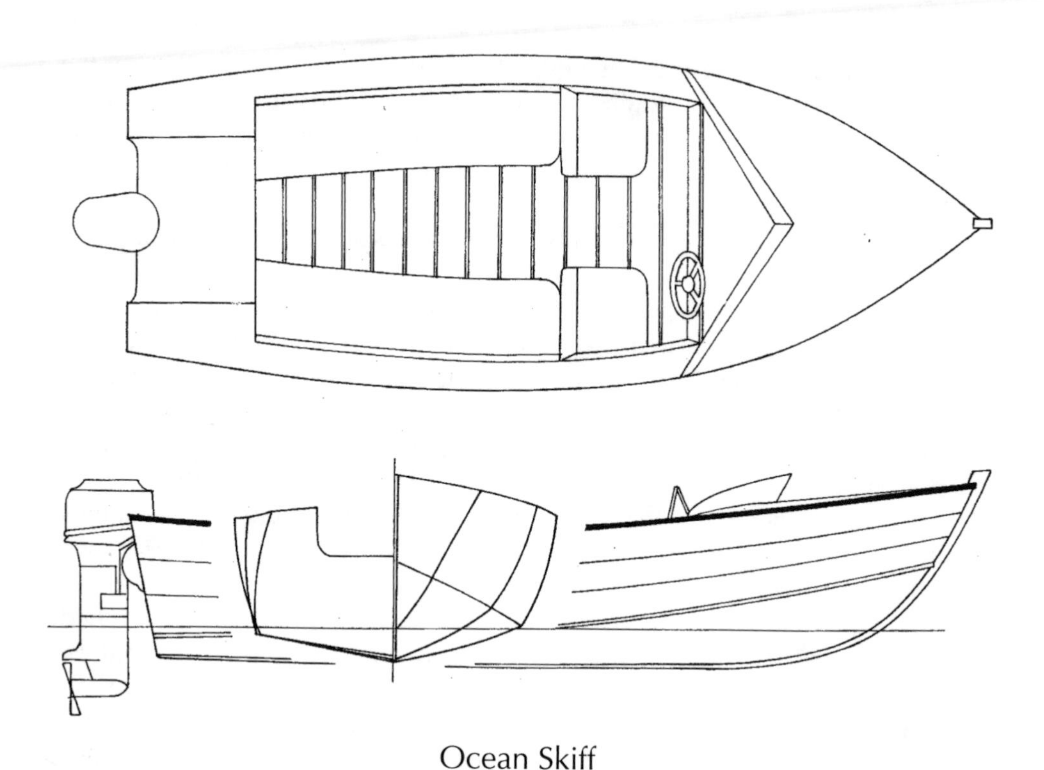

Ocean Skiff

# Ocean Skiff

Tony Hitschler really did want his new runabout to plane. He had an old one that did—Thousand Islander was the model name—built in the days of mass-produced plywood boats. Every summer for many years he trailered her north to New Hampshire. Trailering is hard on even the sturdiest boats, and perhaps Tony's was lightly built. She endured about forty years of it before the inevitable summer when he rolled her off the trailer into Squam Lake and she just kept going down. He dragged her ashore, pried off the hardware, and cut her up for firewood.

He bought a used aluminum skiff, clamped his motor to it, and salvaged what he could of the vacation. He didn't much like the boat—the layout, the noise it made in the water—but he only used it a week or two a year and figured he could put up with it. The other fifty weeks it sat in the garage, largely forgotten. Not until April one year did it occur to him that he might have the wooden runabout he wanted, like the Thousand Islander but nicer. Tony is my nephew, so I was the builder he called.

Tony looks back fondly to all the people, incidents, and even the appliances of his early years. I don't remember his childhood or youth as being idyllic, but the way we think of these things is more important than the way they were. At the same time that he was talking to me about a runabout he was thinking of buying a Volkswagen Beetle, like the one he used to have. It is he who organizes the annual vacation of his sisters and cousins to the resort where their parents took them when young.

The resort imposes a 25-horsepower limit on boat motors, although on the rest of Squam Lake anything goes, as Henry Fonda well demonstrated in the movie *On Golden Pond,* tooling around in an old Chris-Craft inboard and eventually piling it up. The horsepower limit imposed a design problem, because even the most efficient bottom only planes 1,250 pounds, and Tony loves to load up a boat with family. Even if the boat were gossamer, planing in all conditions wouldn't be likely. Time was the biggest design problem: he wanted the boat for the end of July, three months away, and as he lived in Baltimore we had to come to agreement by mail and telephone and pretty quick, too.

I asked him to make a drawing of the boat he had in mind and also to send photos of runabouts that he admired. Taken from a newspaper supplement, the photos differed from those in *Real Runabouts* only in quality of reproduction, and I had to tell him that they weren't three-month jobs. With the drawing Tony did very well. He did not attempt fair curves or other drafting niceties, but he laid out the interior—probably quite like that of the Thousand Islander—with a good understanding of the space each human would take up. Tank, battery, controls: all were considered, and nothing was outlandish or impossible to execute. One whole big phase of the design was already settled!

For the other phase, hull shape and construction technique, I recommended the 15-foot Ocean Skiff of Robert W. Stephens. I could have drawn one myself, but considering what the stock plans cost, I would have been earning less than a dollar for each hour at the drafting board, and July would have been drawing closer day by day. A warped V-shape like the Ocean Skiff's needs considerable development of the bottom sections, which are a portion of the surface of a cone. I have a copy of a very old magazine article that gives a good starting point for locating the apex of the cone, and *Skene's* shows a hull whose topsides are a combination of three cones and two cylinders. But this kind of work takes time, even for people who do it often, with much gritting of teeth for erasures and fresh starts. Stephens gives the curves of the bottom sections with four buttock lines and their offsets, and although I always loft a boat and hang the lofting on the wall until the boat goes out of the shop, Stephens's drafting is so good that it may not have been necessary.

The Ocean Skiff is a model of modern, straightforward, plywood construction, meticulously thought out and drawn. Stephens has quite a range of such boats, power and sail, but he no longer promotes them as he once did. More recently he has been working at Brooklin Boatyard in Maine, and since Joel White's death he has become chief designer. That's a good fit because the fancy White boats always were old-fashioned, and so are the simpler plywood ones bearing Stephens's own name. However, an old-fashioned sailboat is likely to be tippy and slow, but an old-fashioned motorboat is probably more efficient than a current model, and so it proved with the Ocean Skiff.

Thousand Islander looks suited Tony just fine, but the Stephens design takes advantage of what epoxy glue can do, and its greater beam and freeboard (the clinker plywood topsides help to disguise the freeboard) give nearly twice the interior volume of older 15-footers. I made a few minor changes: inserting Tony's layout and substituting Tony's veed coaming for the windshield that Stephens drew for his cold Maine waters.

A more major change was the sectional curve of 4-foot radius that I put into the topsides. Stephens drew the topsides flat, perhaps thinking that this shape would cut down on planking bevels and make the work easier for amateur builders. However, looks are a big part of a boat like this, and slab-sided clinker doesn't look like much. The 4-foot radius was chosen on the lofting (by the way, I now do all my lofting with the boards on the strongback, not on the floor. If you're going to use a strongback anyway, why squat for this stage of the work?), rather than some other radius, to allow the lowest plank to be plumb at the transom and the tumblehome to be what Stephens shows. I notched the frames, which isn't much trouble with only three planks and makes a more solid boat that is easier to keep clean.

Another change that I made while lofting, again for looks alone, was to lower the stern at the sheer by $^3/_4$ inch. By Memorial Day the boat was in frame, and Carol and I walked over to a neighbor's house for the annual beer fest. There we ran into John Binginheimer, already well started on the beer, and I showed him a Stephens sketch from my wallet. He asked to see the boat, and we walked back together. The sketch was now back in my wallet, but John hadn't been in the shop a minute before he said, "There's something different about the sheer aft." Nice eye, John!

When Tony's boat was built, I could still get khaya plywood, whose weight is halfway between okoume and sapele. It is as durable as sapele and its grain pattern swirls luxuriously, which was important because Tony wanted everything but the bottom of the boat to be varnished. That simplified painting because there would be less cutting in, but it meant that the screw heads really ought to be plugged, not puttied, and there were plenty of screw heads because I screwed the ply laps as well as decks and furniture.

Plug grain is supposed to run the same direction as plank grain, but that's easier done with big plugs than little ones, because if the plug head gets covered with epoxy while gluing the sides and bottom, it's hard to see which way the grain runs. You do the best you can. I chop plugs off with a chisel, a little at a time to see how the in-and-out grain is running, and finish them off with a sander. It's not much more work than puttying because wood sands easier than epoxy, and a wood plug stands proud of the surface by itself, while epoxy must be mounded to stand proud.

Khaya is an African mahogany, and I had no boards of it for making plugs, but a fair color match can be made with some kinds of Honduras or with the old lauan that they used to sell. It's always fun after boring the plugs in the drillpress to run the lumber through the bandsaw and see the plugs pop out like magic. However, screw heads must be set deeper for plugging than for puttying, so $^3/_8$-inch stock is about the thinnest that should be plugged.

Stephens shows a plywood sole, boxed in to make a big flotation tank in the event of capsize or holing. But wood does float, and a couple of cubic feet of foam neutralizes the motor's weight. Unventilated places should be avoided in wooden boats because they are the first place for rot to start. In addition, plywood is slick under foot, so must be painted with anti-skid compound. I prefer soles of unpainted white cedar boards, which have some skid resistance even when wet and are very pretty to see. They gray eventually but can be scrubbed, and when scrubbing won't do, they can be sanded.

The Ocean Skiff weighs 425 pounds, but Carol and I weighed her with a bathroom scale that reads only to 350 pounds. Experimenting with a jack, we found the fore-and-aft center of gravity on the keel. We then put three marks on a plank, each 4 feet apart. The center mark went under the keel, the far mark on a block on the far side of the boat, and the near mark sitting on the jack on the scale. We jacked until the boat came off her cradles, noted the scale reading, jacked her down, subtracted the scale reading of jack and plank, and doubled the difference. More recently we have weighed our 2,163-pound *Dandy* catamaran with the same scale: jacking up one hull, but with a plank of 8 feet between keel and scale and 2 feet between keel and block. We could weigh a tugboat with this method, if we could find a meaty enough plank.

The engine remained a question, and I think that if the whole of Squam Lake had a 25-horsepower limit, Tony would have been more willing to abide by it. Stephens rates the boat for up to 40 horsepower, and we toyed with the idea of a bigger engine with cowling decals of a smaller one. An Evinrude sales rep told me that such shenanigans are so often practiced on their engines, which sometimes only vary from each other in manifolds and carburetion, that marine police on many horsepower-restricted lakes carry around with them the parts numbers of the possibilities. Would Tony's Squam resort be that savvy? But in the end he bought a 25-horsepower.

Four of us tried out the Ocean Skiff on the Tuckahoe River in good time before the scheduled New Hampshire rendezvous. In addition to the 425-pound boat, there were 200 pounds of engine, tank, and battery and perhaps 600 pounds of crew, which is perilously close to the limit that 25 horsepower lifts. But lift us it did, without the old dodge of walking aft and then walking forward again. The bow came up, much to the detriment of visibility, but when enough speed had been gained it came down again. After that the boat didn't seem inclined to go much faster, and I'm sure we were doing less than 20 miles per hour. The propeller was the least pitched one that Yamaha sells for that model, and with more pitch I doubt that the engine would have reached peak horsepower.

Still, the warped V-bottom proved nearly as efficient as a flat bottom, and no doubt it would hammer its crew less in a chop, although the wind was calm that day. At the transom the deadrise is only 7 degrees. The "spray rails," as Stephens calls them, probably make a great difference. They are $^3/_4$-inch-by-1-inch strips running along the outsides of the chines, and they look so vulnerable to damage

Ocean Skiff on Squam Lake, New Hampshire
(Photo courtesy Jill Hitschler)

from flotsam and even from trailer rollers that I was wondering if they couldn't be left off. However, I had recently seen John Yank retrofitting a larger version of such rails to his 800-horsepower 60-foot headboats, so went down to ask him about them.

"They ain't for spray," he said. "They give lift and more control too. We're getting a knot and a half, two knots more speed with them rails on." His problem was that in high-speed turns or in coming down off a wave, them rails tended to tear away despite very many throughbolts and gobs of epoxy glass. They should have been incorporated into his mold, which would then have needed to be two-piece to release the hull. But so far Tony's 25 horsepower hasn't pulled any rails off. The job they do is to keep the lifting pressure of the water from escaping out the sides of the hull.

At top speed (and boats like this inevitably are driven at top speed, even when the engine is brand new) a little spray was constantly coming forward from the Ocean Skiff's transom and into the engine well. Probably the motor leg deflected it. It ran out the well scuppers quickly enough, but without a well we'd have had a few quarts of water in the boat at the end of the hour-long trial run. That made me uneasy, and so did the way she steered. The conventional push-pull cable was stiff and lifeless, though there were no hard bends in it. A brake is sold for this system, but it's hard to imagine who would need it. And even the mildest turn put the boat into a skid of sorts. Planing boats do steer that way. I didn't like the wake we were making, especially when I watched it lap over the spartina grass and rumble toward muskrat houses. I don't like planing boats.

Tony was pleased as Punch. The drumming of the water on the wooden bottom he found soothing, and probably it was, compared to aluminum at the same speed. The steering he found perfectly satisfactory, and he remarked more than once how quiet the engine sounded. Maybe it was, compared to his old one; but we're used to four-stroke outboards, and it didn't sound quiet to us.

The Ocean Skiff was a family sensation at Squam Lake later that July. Tony brought home a photo showing eleven people in it, although they sure weren't planing. It's true that the boat is spacious for a 15-foot runabout. It's stable and safe and dry, at least where the crew sits. It has some style and doesn't waste as much horsepower as some boats do. There are a good many worse designs on the market, and among mass-produced runabouts nearly everything is worse.

## Speed King

Five years after *Minnow* in her new hull was launched, I freshened her up for Sam Evans: new paint, decks sanded and oiled, and so on. I don't like such work and tried to get out of it, but the guy I recommended for the job simply didn't do it and robbed Sam blind in the process, so I had to make amends.

The next fall Sam was talking about a new motorboat, this time for his young grandsons who often vacationed with him in Avalon. He and Janet had bought a 9-foot Fatty Knees sailing dinghy, a very expensive production boat. They tried it out a few times, but neither knew much about sailing, and so it sat until Sam got the idea that the grandsons could be crabbing from it. He bought a 3-horsepower outboard, but when the power was applied the boat just sat up on her fatty posterior like Emil's flatiron with 10 horsepower but very much worse, being only 9 feet long and with a wineglass transom. The boys couldn't see out of her, and there was no point telling them to throttle down.

I suggested a garvey because I like the boat and because the New Jersey tradition of garveys would appeal to Sam. "Everything is international so nothing has any meaning," he once said to me in another context. The trouble with garveys is that for efficiency and for directional stability they need to be long and narrow, and to make a garvey wide enough for the kids to run around in, it would be longer than *Minnow.* That was impossible, Sam said. Grandpa couldn't have the shortest boat in the fleet.

I mulled Sam's needs over, and it seemed clear a 3-horsepower motor would drive whatever we put it on at hull speed and wouldn't drive anything faster, so efficiency was not paramount. For directional stability perhaps a couple of 1 $^{1}/_{2}$ -

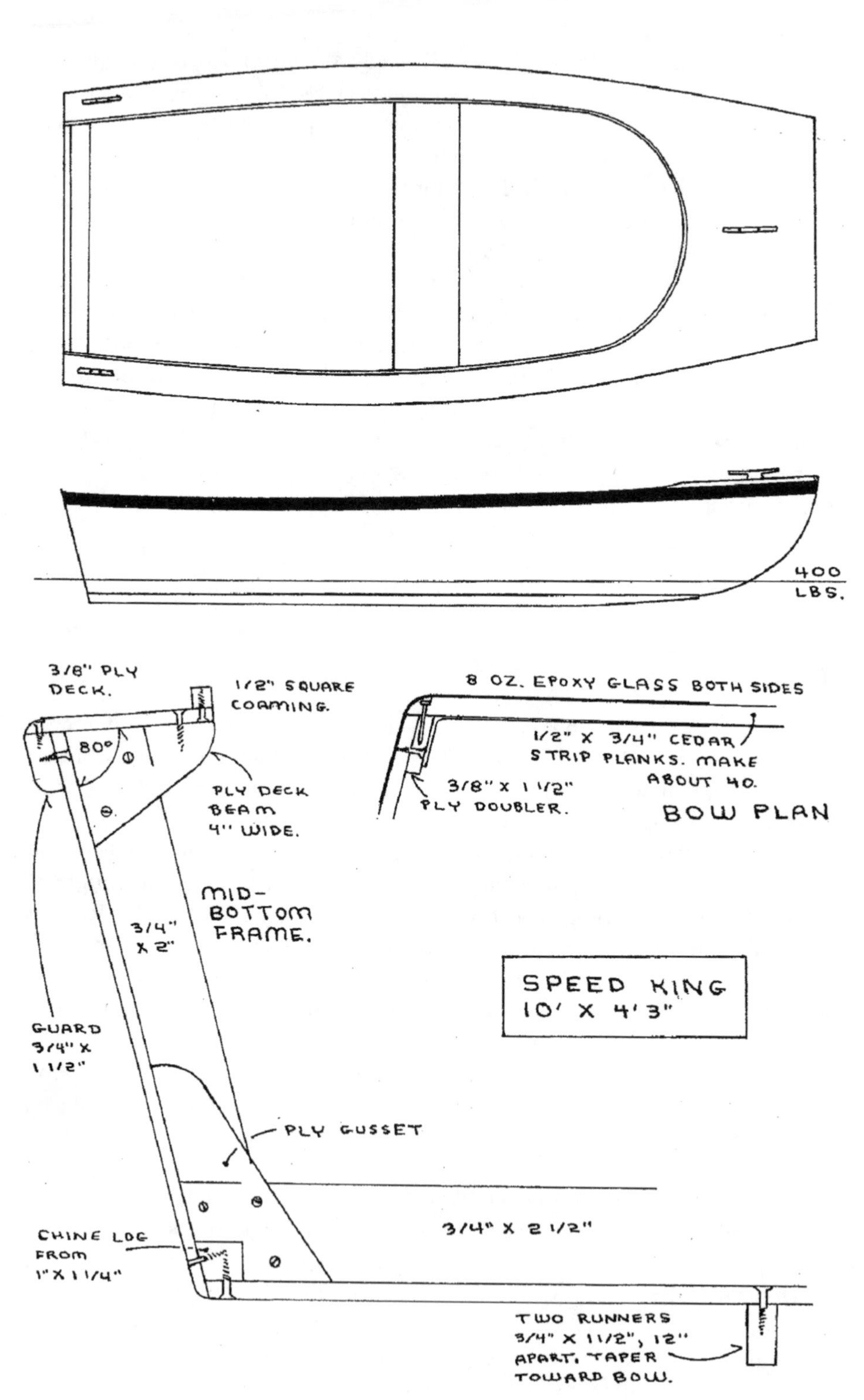


*Speed King* with details

inch-by-$^3/_4$-inch runners on the bottom of a garvey might substitute for a more normal length-to-beam ratio and the more gradual chine curves that would go with it. The runners would also allow a sole uncluttered except by a single frame. Considering the amount of activity that goes with crabbing (not to mention youth), that seemed a good idea. Certainly a short fat boat would weigh less than a longer one, which mattered because Sam planned to launch it when the kids arrived and haul it when they went home.

I sent him a study plan, and he thought it would work. "The fat garvey concept is fine with me," he said. He wanted a bunch of bronze hardware—oar locks, flag pole, running lights—which I persuaded him weren't needed for the very constricted waters where the boat would be used; but I did stick a huge bronze handle on the bow instead of a cleat, which is helpful when picking the boat out of the water and also gives a look of studied magnificence. I had had it for twenty years, having bought it from a cobwebbed chandlery where it may have lain for fifty years, waiting for J. P. Morgan or his ilk.

I put in a full-length thwart stringer about 10 inches off the sole, and you could do the same if building one of these boats. It isn't needed for topsides stiffness, but it does allow thwarts to be put in experimentally and moved if they don't suit. In planked garveys such stringers are often seen, and typically they have dozens of screw holes in them where the owner tried a thwart and then thought better of it. But in a garvey this small I would settle for one lift-out thwart sitting on a short glued-in stringer. Its aft edge would touch the midships frame, and it would mostly be left in, to minimize stubbed toes on the frame.

Sam wanted stern sheets, which you can see in the photo. These U-shaped seats are more often seen in traditional rowboats, but here they and the deck are bored for fishing-rod holders. In a boat this light it isn't possible to sit aft beside the motor without heeling alarmingly, unless another person on the forward thwart trims ship. There is a picture of Sam's 200-pound son-in-law sitting beside the engine in *Speed King,* and she looks close to capsize.

The boat is best steered from the forward thwart with a piece of PVC pipe slit and hose-clamped over the tiller, as used by the garvey club. Even the cable steering used on *Puxe* is too complex for this boat. As boats get smaller they *must* get simpler, and this one is certainly too small for the electrical system that would power running lights, especially as 3-horsepower motors don't generate electricity. If one ever wanted to take the boat out after dark, a dry-cell running light attached to the deck with a suction cup would be the ticket.

Sam specified okoume plywood rather than sapele for the 30 pounds it would save, and that was a good choice. The finished boat weighs just over 100 pounds, which is too much for cartopping but easy enough for two people to pull out of the water and turn over. As designed she will take up to a 10-horsepower engine, and I was thinking that when the grandsons were old enough to have some sense as well as some impatience, they might rather plane her at 15 miles per hour than putt along at 4 $^1/_2$. But if the motor and its fuel weigh 100 pounds, then even with 10 horsepower this garvey will only plane 300 pounds of people. I hope that when these boys are grown they won't be as husky as their father.

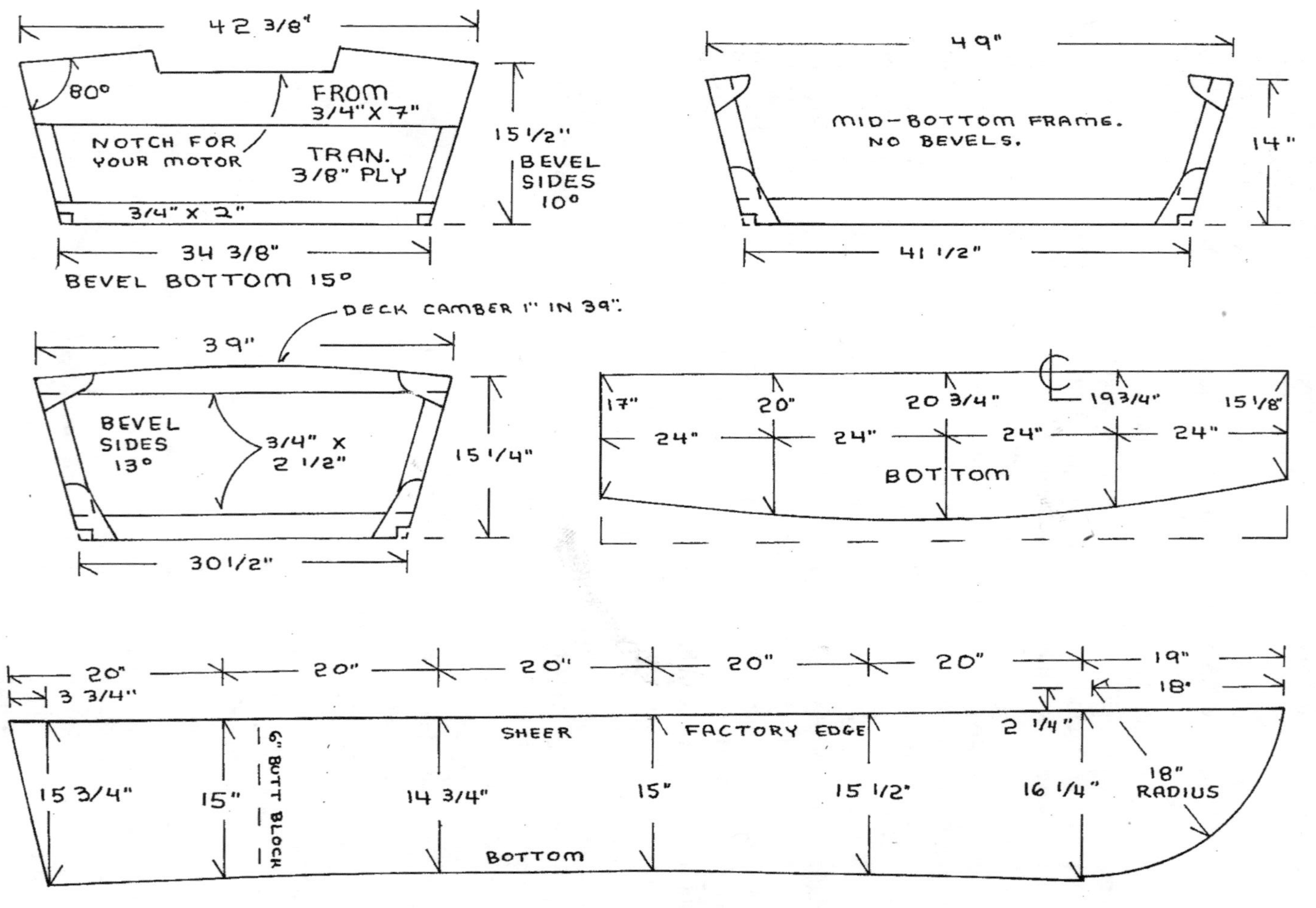

*Speed King* frames and ply cutting

*Speed King*

Phil Bolger saw *Speed King* just before delivery and said that although the runners posed no problem for 3 horsepower, with 10 they would create turbulence that might keep the boat from planing. He suggested a wooden wedge between the runners: the kind of transom wedge that people sometimes put onto over-heavy powerboats, hoping for the planing miracle. I'm sure Phil is right about this.

*Speed King*'s plans, like *Chingadera*'s, are for building without a strongback, as many people like to do these days. She needs one sheet of $^{1}/_{2}$-inch ply and two of $^{3}/_{8}$ inch and a few feet of Douglas fir for the framing and cedar for the thwarts. I usually figure 100 screws per sheet of ply, using $^{3}/_{4}$ inch #6 for $^{1}/_{4}$-inch ply, 1 inch #8 for $^{3}/_{8}$-inch, and 1 $^{1}/_{4}$ inch #10 for $^{1}/_{2}$-inch. She's a simple boat to build, but that's no reason to build her of inferior materials. A friend of ours once built a tortured-ply kayak, and after the trouble of cutting out the hull panels and wiring and taping and sanding the glass, he put a sheer clamp into her that was ripped from framing lumber and liberally sprinkled with knots. The sheer in plan was then a series of flats, not a curve, and although he put on a deck and finished the boat, he has never dared take it out where other people might see it. The smaller the boat you're building, the less painful it is to use good materials.

Cut out the bottom and the topsides first, with the saw set at 15 degrees. The scrap left over can then be allocated for deck, transom, and gussets for the mid-bottom and forward frames. Assemble the hull first without glue. Screw the bottom runners on. Screw the chine logs and the frames to the bottom, and then screw on the topsides. This is all best done right side up on a couple of sawhorses. When you see that the joins are good, take it all apart and glue it together.

The bow then needs to be pulled in with a Spanish windlass, as described in the section on the mini-garvey in chapter 2 and shown in the photo. Holes for the windlass should be drilled in about the same places, and the bows pulled together until the planking at the sheer is 32 inches apart, inside-to-inside. The ply doubling for bronze nails or other fastenings can now be added, and the bow edges beveled off for strip planking.

Tighter fits can be made if one edge of each strip plank is beveled off to 3 degrees. Keeping these strips aligned with each other is easier than it was with the mini-garvey, because $^1/_2$-inch stock is thick enough for drilling a hole and driving a nail through two strips wherever needed. This strip planking is messy work, and after each session the tools need wiping down with acetone or other solvent. Take care to clean up the inside of the bow with a putty knife as you go along. Glue blobs on the outside of the curve are easy enough to sand off later when making a smooth surface for glassing, but on the inside it's harder.

When glassing over wood, it's important to get all sanding dust as well as larger particles off the work before laying down and tailoring the glass. Most fiberglass cloth has a whitish binder that dissolves in polyester resin, which makes the cloth stiff as well as less prone to fraying. If you're using a cloth with binder, the cloth has to be gored to go around the corners. If you're using a cloth without binder, meant for epoxy resin, it will stretch and conform around the corners without being gored; but it will stretch in other ways too, with the whole cloth changing dimension as you work the resin into it. The cloth with binder seems best to me.

For the sheer, a single longitudinal 1 inch by $^3/_4$ inch does double duty as guard rail and sheer clamp. A doubler inside the bow gives a better landing for the fastenings there. The deck is best screwed down without cutting the semicircle for the coaming. Then the 18 $^1/_2$-inch radius is drawn, and temporary blocks are screwed on inside it to make a form for the laminations. Four layers of $^1/_8$-inch stock will do it, and if you are brave and have enough clamps you can glue them down at the same time they're laminated.

The two jobs on this garvey that require more thought and labor than building the simplest johnboat are strip planking the bow and making the round forward coaming. They are also what give the boat some distinctiveness and make it a pleasure to use and a boat that other people admire. A shirt that is made out of a flour sack, with holes cut for the head and arms, unquestionably fulfills the function of a shirt. But in news photos that we see from very poor countries, if people have shirts at all they're likely to be cut to fit the body, and they're likely to have some ornamentation, even if it's just the logo of a commercial product. If these people can afford more than a flour sack, we can afford more than a johnboat. And from a practical point of view, the square spoon bow of a garvey knocks down spray from small waves better than any bow I know. As for the coaming, doesn't it look pretty?

Sam stopped in from time to time as the garvey was a-building. He wanted a name on the transom, he said, in the same style as *Minnow*'s. For that I called in a sign painter who is clever at making yellow paint look like gold leaf. *Speed King* was the name to be painted, and Sam explained it this way:

He owns a strip of beach in Avalon, and one day he saw two men approaching in a motorboat. When they reached shallow water one man jumped out, bringing with him a pair of water skis and one end of a tow rope. The man still in the boat began backing up, with the other end of the rope attached to the transom. The man on the beach was putting on the skis. Sam thought he'd better go down.

"This is my beach," he said to the man wearing the skis. "You're welcome to use it. But that isn't the way people usually start out on water skis."

The man eyed him suspiciously and took a firm grip on the handle of the rope. "It's the way we always do it," he said. He turned his head toward the boat and shouted, "Let'er go, Speed King!" Sam figured that as long as it was going to happen, he might as well stand back and watch.

I delivered Sam's *Speed King* in the spring, and by summer he was able to report that she stayed pretty level no matter where the boys were in her and no matter how hard the throttle was twisted. Their father also seems to find her satisfactory for fishing. There haven't been any maintenance problems. That was four years ago, and I'm in touch with Sam, but haven't heard any talk about moving up to 10 horsepower. Perhaps the boys are still too little, or perhaps Sam is cautious. At any rate, it's nice to know that the potential is there.

# 4

# *Long, Narrow* *Powerboats*

The long, narrow powerboat is a romantic relic from the days when engines were too heavy to lift boats out of the water. But they excite our admiration more than other relics, such as carpet sweepers whose brushes are geared to turn contrary to the wheels. Such boats are something like riding horses: when we see them from our automobiles, we can't help thinking that we're going faster—but those riders, they're onto something we're missing. *That's* the way to get places! A long, narrow boat that's in the water has a softer ride than a fat one that's planing because the waves aren't banging on its bottom. It also consumes a quarter of the fuel of a planing hull at the same speed. But its grace and attractiveness are all out of proportion to these practical considerations.

As discussed in the chapter on runabouts, narrow length relative to beam makes a more efficient hull, even if it's a planing hull. In a recent issue of *Professional Boatbuilder* magazine Steve Callahan discusses the narrow powerboats of designer Adrian Thompson, best known for his racing multihull sailboats. Thompson has interested the British military in his patented "very slender vessels," which have a length-to-beam ratio of about 6:1. They do 50 knots and more, but their narrow bows are wave piercing, which eases the motion but is incredibly wet, so their future use as pleasure boats is uncertain. But even for boats that bounce along on top of the water, giving Granny a blurred sightseeing trip and aggravating her piles, the ones that win races are very narrow compared to conventional runabouts.

Like Adrian Thompson, Captain Nat Herreshoff is best known today for his racing sailboats, which won the America's Cup countless times; but he probably made more money in the first half of his career building powerboats, typically with length-to-beam ratios approaching 10:1. One hundred feet or more in length and steam-powered, these boats would do 25 or 30 knots without squatting, let alone planing. Herreshoff's powerboat downfall came with the development of internal combustion engines. He was accustomed to building steam engines in his own shop, which was self-sufficient to a degree hardly imaginable today: whether making sails, rope, hardware, engines, or the very power that ran the shop tools, he bought nothing but raw materials from the world outside. He stuck with steamers, and he was building them as late as 1920, although the list of customers shrank. Steam engines not only are heavier than internal combustion engines, but they use three times the fuel to make each horsepower.

Twenty years ago Weston Farmer gave us a wonderful book, *From My Old Boat Shop*. He tells us a great deal about long, narrow powerboats, but he also admits that shorter, lighter boats eventually took over the market. The ride of these early planing powerboats was just awful, and ninety years later there is no better fix than a great deal of deadrise with its concomitant loss of efficiency and increase of wake. So for most of us, the usual weekend nautical excursion has us slowing down for bigger boats and shaking our heads at their poor manners and then blasting across the bows of smaller ones, which are forced to slow down for us.

Farmer gives us the lines of *Coyote II*, round-bilged and 25 feet by 6 feet. He drew her a foot wider than the original *Coyote* designed by Edson B. Schock in 1907 in which you might have to "part your hair in the middle." With her rocker and relative lack of lifting surface aft, she may not look at all like a boat that would exceed hull speed, which would be 6.7 knots. Farmer prints a letter from Schock's son saying that a replica of the original *Coyote* built in 1946 with 25 horsepower inboard was timed at 16.3 knots. At that speed-to-length ratio the boat should have been planing, but *Coyote* with fuel and crew likely weighed 2,000 pounds, and a horsepower won't lift 80 pounds out of the water, no matter what the shape. What's happening here?

Hull lines were once the subject of much study. Somewhere I once read a haunting sentence about lonely old men in public libraries studying hull lines in back issues of yachting magazines. In Captain Nat's day he and many other designers derived the lines from models that they carved. You didn't just see the lines; you felt the water flow with your hands. That allowed Captain Nat's blind older brother, John Brown Herreshoff, to design boats also. However, whittling was not the inevitable designing method in times past: in Frederik Henrik Chapman's treatise on shipbuilding, *Architectura Navalis Mercatoria*, published in 1768, hull lines are drawn. Tedious work it must have been before Isaac Newton's invention of calculus made possible the curves of diminishing radius that we all use for drawing lines today. Previous to that, sectional lines were drawn with a compass set to many different radii, all tangent to each other and all in drippy India ink, not the nice felt-tipped pens that we have now.

Too much can be made of hull lines and was by Nils Lucander, who died only recently. For many years he peppered the yachting press with letters explaining how his boats used special shapes to sustain speeds above hull speed. A fat cruising ketch was happy for hours at a speed-to-length ratio of 1.7. I doubt that many people took his theories seriously. Today numbers have supplanted lines as the primary matter in boat design. The most important number is weight, although length, power, wetted surface, and prismatic coefficient are very important too, as are various ratios. Lines are seen more as a way to achieve these numbers than as an independent entity.

Sailboats are like the traditional long, narrow powerboats in that power is limited, so resistance really matters. In *The Sailing Yacht* Juan Baader tells us that when a Dragon—which used to be an Olympic class—is sailing to windward at 4 knots, the wave-making that is the result of her lines is only 30 percent of her total resistance. The other 70 percent is almost equally divided between water friction (wetted surface) and the wind resistance of her hull and rig. Of course, a sailboat has more wind resistance than any powerboat short of a houseboat; but when we take our low, narrow *Puxe* down the Tuckahoe River in windy conditions, she speeds up noticeably downwind and slows down when a river bend changes her course 180 degrees.

This chapter shows two long, narrow powerboats. *Puxe* was designed with numbers as my first thought, but the Buy-boat is an adaptation of a design in which lines were considered most important. The Buy-boat's plans are shown here, but no example has yet been built, so you must judge for yourself how well you think it would work.

# Puxe

*Puxe* is 22 feet by 4 feet, 10 inches overall and 21 feet by 4 feet, 3 inches on the waterline, making her length-to-beam ratio 5:1. She is of composite build with a fiberglass foam sandwich bottom, clinker cedar topsides, and plywood deck. She is the successor to a boat of the same name and dimensions that was built of cedar. We had that one thirteen years, and her demise was as interesting as her life.

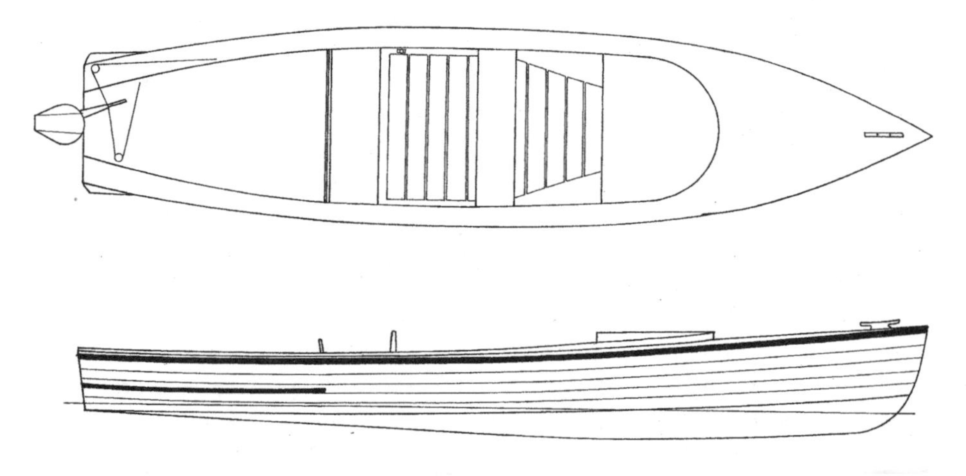

*Puxe*

The original *Puxe* was built the first winter after we moved to Tuckahoe, when I was infatuated with the materials and methods of the old-time South Jersey builders. In the days when sailing garveys were the usual workboats they were built of white cedar boards, so that's what *Puxe* would be. A garvey's topsides were cleated together and bent around a mold or two and the bottom was then whacked on, so that would be my method. How I justified using cross-strip planking and modern glue for the bottom I don't remember. Why I didn't work in a hardwood chine log and fasten the topsides and bottom to it I also don't remember, but the cedar-to-cedar join in this crucial spot eventually cost us the boat.

White cedar is not a bad boatbuilding material, and it must have been better when it came from virgin forests—or should we say virgin swamps? That wood was denser and had fewer knots. Logs were big enough to make quarter-sawing feasible, so the growth rings were at right angles to the face of the boards, which swelled and shrank less. Today's white cedar is still more dimensionally stable than some woods, and the heartwood is as durable as teak. The core or pith of the log must be cut to waste along with the sapwood because it too tends to rot, and it too readily twists or cups a board. But cedar works easily and is not likely to split or splinter. It is light weight, at least until it's put in the water. Like cypress, immersed cedar soaks up its own weight in water and is then heavier than soaked-up mahogany.

The glue used in the first *Puxe* was resorcinol, and it's still a good alternative to epoxy, especially for the people who develop an epoxy allergy (and there are quite a few who do). Fits must be closer, and it stains the wood badly with a color that bleeds through latex paint; but it is less toxic to work with than epoxy, and the pot life is longer. It absolutely is waterproof, and over time plastic resin glue and Airolite are not. Resorcinol can't be used for filleting or fiberglassing, and it's less readily available than epoxy, whose clamorous advertising has swept the market nearly clean.

Epoxies are less different from each other than salesmen would have us believe. They are something like paint: the manufacturers of the resins and pigments used in paint will tell you everything about them and will even provide you formulas for making paints of various qualities. The people who mix and market paint would rather throw sand in your eyes. *Their* paint contains secret ingredients. *Their* formula is superior. *Their* product lasts longer, is glossier, penetrates wood more deeply. It's the same with epoxy, except that most of the vendors don't even mix their own products; they buy them wholesale, package them, and market them. Good paint lasts longer than cheap paint, and it's a bargain, because it takes a great many hours to put on a few dollars worth of paint. That is even truer of varnish. But I have not noticed any difference in performance between industrial epoxies and the highly advertised brands.

Because we were interested in comparing old *Puxe* to the long, narrow boats described by Weston Farmer, we timed a number of early runs, taking account of the tides, of flat water and choppy, of weight aboard and the weight of the wind. With a 10-horsepower outboard and an all-up weight of 900 pounds, she cruised at somewhere between 9 and 10 knots and could be pushed to make a couple of knots more, although that speed produced a good deal more engine noise and may have doubled fuel consumption. That's a speed-to-length ratio of just over 2. We haven't timed the composite *Puxe* as carefully, but we believe she's a knot or two faster with the same engine, which we attribute to lighter weight and perhaps to better lines. Old *Puxe* weighed 450 pounds when launched, but soon soaked up another 100 pounds of water. New *Puxe* weighs 325 pounds, and although fiberglass does soak up water to some extent (where else would osmosis come from?), the extra weight isn't significant.

Old *Puxe* stayed tight for ten years and thousands of hours of use, and she outlasted her first engine, which we replaced with a 7 $^{1}/_{2}$ horsepower of the same displacement. Both were Honda four-strokes, and the only difference we can detect in them is the price, although the 10-horsepower might push harder at maximum revolutions per minute. Many engine manufacturers offer two or even three horsepower ratings for the same displacement engine, and unless you are planning to race the boat, the lowest-rated one is the best value and may well last longer, being less stressed.

Eventually wakes and other chop started flexing the chines of old *Puxe,* and she began to weep through them. The wet wasn't cataclysmic, although it can be. A childhood friend recently told me that when he was in high school he crewed for a whole summer on a racing plywood sailboat—perhaps a Moth, although he didn't remember—and on the morning of the last regatta the skipper told him that they would win the season's cup wherever they placed that day. They had only to finish. But on the windward leg they tacked and stomped too hard on the bottom, which detached itself from the topsides, and suddenly they were swimming. The Moth was a development class, and many of them were built too light.

Old *Puxe* didn't do anything that dramatic, but no one likes a leaky boat. For the next couple of seasons we pumped her only when setting out, but we could foresee the time when she'd need pumping more often. We weren't going to put up with that as long as Al Whitehead did with his old plywood daysailer. I made a half-hearted attempt to sell her, but wasn't willing to lie about her problem, so found no buyer. To the dump she went. (For cutting up a wooden hull a circular saw with handsaw back-up is best. A chainsaw blade takes more offense to bronze fastenings than does a carbide circular-saw blade.)

The foam-sandwich bottom of new *Puxe* allowed me to make her any shape I wanted. With cross-planked bottom, the sections of the old boat could only be straight lines. When its plans appeared in my first boating book I had many letters asking for a plywood version, but never drew it because the conical projection would have made the forward sections convex and the entry too blunt. Then we visited the motorboat museum in Clayton, New York, and saw there several vintage launches with hollow sections forward and very fine entries. Building with foam sandwich, it wouldn't be hard to imitate them.

The old hull had rocker, like *Coyote,* but in the new one most of it came out. That allowed a higher prismatic coefficient (in other words, more volume in the ends), which made up for the reduction in prismatic of the hollow forward sections. Prismatic is a ratio comparing the volume of underwater shape to the vol-

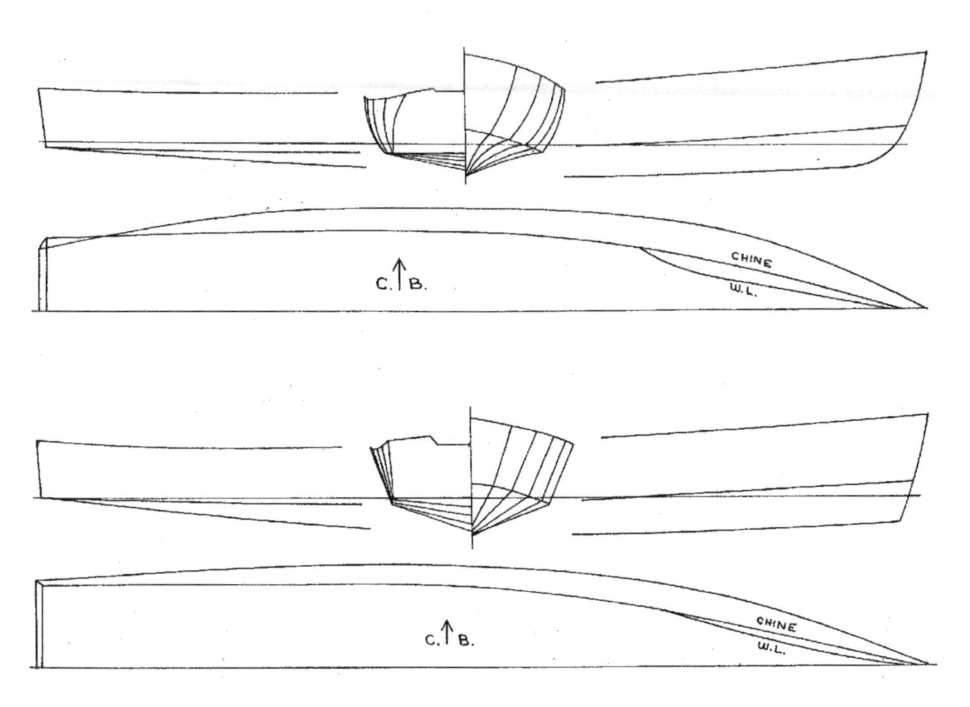


Lines of old *Puxe* (below) and new *Puxe* (above)

ume that would be there if all sections from bow to stern were identical to the midships section. It is written as a decimal. A very fine-ended hull might have a prismatic of less than .50, while a houseboat with two plumb transoms would have a prismatic of 1.00.

Every speed-to-length ratio has an ideal prismatic. At 5 knots a 25-foot waterline hull should have .50, but at 10 knots the ideal is .70. That's hard to arrange in sailboats, which move at different speeds depending on wind strength, and even in powerboats some thought to angle of entry or even looks may compromise the prismatic. Within limits, overladen boats go better than underladen ones, because overloading just pushes the hull down into the water a trifle, while underloading shortens the waterline on which speed depends. Similarly, too high a prismatic hurts performance less than one too low.

Foam cores are all the rage in fiberglass boats, although only the most expensive of mass-produced models use them, because the foam itself is expensive. The benefit of foam is that fiberglass itself is strong but not stiff, meaning that while thin fiberglass withstands local impact it does not hold its shape, especially if the shape is flat. Early fiberglass boats solved that problem by making the glass thick—sometimes as thick as the wood of the hull being copied—but the weight was then ruinous. At the present stage of technology, a foam sandwich hull that is built without vacuum bagging or exotic laminates is about twice as thick as a solid fiberglass hull and weighs about half as much. It can be built with only a rudimentary mold.

Of the many kinds of foam on the market, the best for one-off building weigh about 6 pounds per cubic foot, are dense enough to hold screw threads, and come in solid sheets, not scored into 1-inch squares (those are for hulls built in a female mold). Although epoxy resin can be used with the glass, few builders find it cost effective, and polyester or one of its sophisticated siblings such as vinylester is thought best. Sewn Fabmat, rather than the old woven Fabmat, is easier to work with and gives a better result, with higher glass-to-resin ratio. Its extra cost is nearly recovered in the smaller quantity of resin needed.

To build the new *Puxe*, wooden frames were set up on a strongback; wood stringers $^5/_8$ inch thick were then screwed to the frames. The foam was screwed to the stringers from underneath with 1-inch sheet-metal screws, and the outside was covered with fiberglass. The first plank of the topsides is solid fiberglass, with the two glass skins joining at the chine. To make a mold for the outside layer I tacked about 5 inches of thin plywood to the frame sides, covered with cellophane as a parting agent, and laid the glass over that, too. The glass cured wavy and unfair, and straightening it out with wood battens before laying up the inside glass was a nuisance. It would have been better to have laid up those 5 inches of fiberglass on a table, as was done with the Dobler, with the skins of outside and inside laid up over it. If those glass strips had been $^1/_8$ inch thick, the weight of the whole boat would have increased by only 20 pounds, and the work would have been neater.

The fiberglass part of the hull—I think of it as a shoe—was removed from the mold and glassed on the inside. To hold the shape of a foam-sandwich hull during this step, it is necessary to use props or perhaps molds temporarily glassed onto the outside because the hull has little stiffness with only the thin outer skin of glass to hold the shape. To fill the weave of the cloth a coarse filler like q-cells is better than the fine fillers used with epoxy resin because it sands more easily. The filler is put on with a trowel. Polyester resin takes many months to cure if exposed to air, so when it is sanded it clogs the paper quickly. The solution is to mix a bit of liquid wax into the last layer of resin, which migrates to the surface while it cures and seals out air. Resins and gelcoats can also be bought with the wax already added, but gelcoat is most useful to production builders who spray it into a mold. One-off fiberglass hulls, or shoes, are best finished with epoxy coatings, which seal out moisture better than gelcoat and give some protection against osmosis.

The topsides of *Puxe* have a radius of 30 inches, arbitrarily chosen and serving no purpose but looks, just like Tony Hitschler's Ocean Skiff. Sawn frames were glued and screwed to the solid fiberglass bottom plank, and the lowest clinker cedar plank was through-bolted to the fiberglass. After planking, more frames were steam-bent in, screwed only to the cedar. The wood-to-fiberglass join has proved perfectly strong and tight, although the strains on it are not great except when the hull is turned over for the winter. Polyurethane stickum was used in the laps, but lately friends have been telling me that polysulfide, although it does not stick as determinedly, remains flexible longer and does not have the annoying and expensive habit of setting up in the tube as soon as the factory seal is broken. Epoxy should not be used to join wood and fiberglass because the two expand and contract at such different rates. It's like fiberglassing a wooden hull: sometimes on a dimensionally stable plywood hull the job lasts, but on a planked hull it always gives trouble.

We do like the new *Puxe*. Being so much lighter than her predecessor, she is less stable to step into, and perhaps because she is lighter she seems smaller, although her dimensions are the same except for the unused space that she loses to her tumblehome aft. She steers with a whipstaff (a vertical tiller) and a simple system of cables and pulleys. The engine kill button is also brought forward to the helmsman's seat, where it is immediately available if we come around a creek bend and find a swimmer—human or animal—directly in front of us. For ordinary use, the kill button is handier in its original place on the engine.

Remote controls can also be bought that bring the throttle and shifter forward, but an open boat really shouldn't need that. In an automobile you change speed often to cope with corners and traffic, and boat skippers often do likewise, for no better reason than they're used to it or are expressing their emotions that way. But powerboats are not automobiles. They are very much slower, and each has a certain speed that it is designed to run. Why not just set the speed and steer as need be? You do not see commercial boat skippers fiddling with the throttle, except when maneuvering in small spaces.

After a couple of seasons we fitted the new *Puxe* with an automatic bilgepump. It's a conventional 12-volt pump with a conventional float switch, powered by two 6-volt lantern batteries wired in series. Many people, including Phil Bolger, doubted that it would work. Susanne Altenburger, who loves to plan mechanical systems, favored a motorcycle battery with solar panel, perhaps augmented by the little alternator that the Honda does have. That would require a good deal more wiring, and as the boat slip faces south, the boat would have to be backed in, or else the panel with its wires would have to be set up on thwart or foredeck and stowed each time we used the boat.

The dry cells last only a season and cost about $15 a pair, but they do the job. The cockpit area is 45 square feet, so each inch of rain brings in 28 gallons of water. Dry cells don't like to be used steadily because they get hot and can expend more of their finite energy in heat than in doing work. But the feeblest marine bilgepump moves 350 gallons per hour, and it seldom rains more than an inch an hour, so our bilgepump batteries are resting most of the time, even in the worst downpour.

A pump like this can deal with all the rain in the sky, but not with seas coming aboard or a hole in the bottom. That requires a larger pump and probably an internal combustion engine to power it, not electricity. Many a boat 22 feet long has a self-bailing cockpit, but that in itself adds weight, and to work well the cockpit needs to be 6 inches above waterline while *Puxe*'s sole is an inch or two below waterline—even the weight of just two crew is still 40 percent of total weight, and in such a narrow hull, they need to sit low.

Another reason for sitting low in *Puxe* is the pleasure of being near the water. A flying bridge is the antithesis. I was in a flying bridge only once and didn't stay long. An airplane flying low over water gives as much sensation of a nautical experience as does a flybridge, and the motion is very much pleasanter. Even in the Ocean Skiff, which does not have a self-bailing cockpit, planing lifts the crew out of the water along with the boat, and that combined with the skitterish steering of a planing hull gives the same sensation that ice skating on stilts might do. Sitting in *Puxe* with the water going by only a few inches away, we feel almost as much a part of it as when swimming.

Another improvement that we made in *Puxe* after a couple of seasons was to bore a 2-inch hole in the foredeck. The bulkhead between cockpit and cuddy misses the bottom by an inch, both for ventilation and to avoid a "hard spot" when the bottom flexes (even foam-sandwich bottoms do flex to some extent). I thought that ventilation would be enough, but the steel gas tank, which is locked in the cuddy as a way of locking the boat, was rusting faster than I could paint it, and the canoe paddles that are our auxiliary power were delaminating. The lifejackets smelled funky. Despite having a white deck to reflect heat, the cuddy needed *cross* ventilation. The rain that comes in the 2-inch hole is negligible compared to the condensation formerly generated. A ventilator on top of the hole would keep rain out but would be in the way when we handled docking lines or set an anchor.

We have had bottom-paint trouble with *Puxe*, whereas on our cruising multihulls the only problem has been that for years we painted too often. Then I noticed that Tom LaMers, who trailers his cruising catamaran and sometimes sails in company with us, appeared never to have painted the bottom of his ten- or fifteen-year-old boat. "What do you use on the bottom?" I asked.

"Once in a while a little steel wool," he said.

Our *Dandy* is in the water only about six weeks a year, but during that time we sail her a thousand or more miles, so water is often going past the hulls. Crossing from the Canaries to Barbados years ago in another catamaran, the tradewinds and the resultant bow waves were so steady that we grew gooseneck barnacles halfway to the sheer; but critters in temperate waters are not so aggressive. Recently I've been painting *Dandy*'s bottoms every other year, and nothing has grown on them. This coming season I'm going to stretch it further, and maybe we'll end up like Tom, with just a little pad of steel wool.

*Puxe* is another matter because she's in the water seven months a year. We always use high-copper-content paint, although the stuff with the highest content is hard to brush on smoothly and can wind up looking like texture paint, which slows the boat. Adding powdered tetracycline to the paint, which was a fad a few years ago, only makes matters worse. However long you stir it, the powder forms little balls when the paint is brushed on, which literally must be sanded off before launching or the hull goes through the water with the resistance of a porcupine.

Our neighbor, from whom we rent *Puxe*'s slip, keeps his own boat on the end and has no bottom-paint problems. I mentioned that when discussing our troubles with a paint salesman and also said that on a blow-out tide *Puxe* takes the mud, and where the mud has touched the bottom is where the barnacles grow—all this in water that is fresh except in severe droughts!

The salesman explained that to work, bottom paint must slough off, and to do so water must go past it. On the end of the dock there is enough tidal current, but it's less at *Puxe*'s stern and nothing at her bow. To make matters worse, we don't take the boat out for the many weeks we're away each summer, when the water is warmest and the critters most active. It would help, the salesman said, if we scrubbed the affected parts with a cocoa brush immediately before leaving home and immediately after we return. The seasons when we remember to do that, *Puxe* has few or no barnacles when she's hauled out in December.

When thoroughly soaked up, the hull of old *Puxe* weighed 225 pounds more than the new one, or 70 percent more. But we never used either boat without 100 pounds of motor and fuel and perhaps 25 pounds of bric-a-brac in the cuddy. We seldom go out solo, so there is usually 300 pounds' worth of crew in her—and often another 300 pounds when we invite another couple. Then the new boat is only 18 percent lighter than the old one. It still does matter to performance, but the biggest difference comes when we roll her over for the winter; with old *Puxe* we had to use levers under one side, but the new one we simply pick up.

The hollow sections forward may not be a good idea, but they do look pretty. The entry is so fine that we may not be using the whole waterline. The photo shows her making about 12 knots. Is that a speed-to-length ratio of 2.6 using a 21-foot waterline or a ratio of 3.0 using a 16-foot waterline? Phil Bolger once designed us a 32-foot-by-8-foot cat yawl, which we did not in the end build because we decided we wanted another catamaran. In the drawings the yawl appears to have a waterline of nearly 32 feet, but Phil said the entry was so fine that it probably would never make a hull speed of the square root of 32 times 1.34.

For sure, taking the rocker out of the stern was a good idea. Old *Puxe* had perfectly straight chines and keel approaching the stern, but at rest her transom just touched the water. At cruising speed she squatted enough to put 2 or 3 inches of her transom under. Who knows whether the boats that Weston Farmer described squatted or not? He said they achieved speed-to-length ratios of 3.0, but in his two photos the boats are going much slower than that. Farmer is more cunning than he would have us believe.

In the new *Puxe* lines drawing, the chines are immersed 2 inches amidships and 2 inches at the stern. Under way she has the same 2 to 3 inches of transom immersed, but she squats noticeably less. She throws even less wake than her older sister did, but her lighter weight may explain that, too. The lines of the new boat do move the center of buoyancy 9 inches farther aft, so it's more directly under the weight of engine and crew.

People notice the Ocean Skiff, with her varnished hull and retro styling; but even people who aren't interested in boats stop and stare when *Puxe* goes by. In looks and in the way she moves through the water, she has nothing in common with other boats they've seen. That's fun for the owners, but it's hardly enough to justify a long, narrow boat. In serpentine creeks her beam allows her to penetrate farther, but her length makes her awkward to turn around. She is not a good boat for chop, let alone genuinely big water. I thought her clinker planking would catch

New *Puxe* at 12 knots

some spray and make her drier, but it doesn't. Half throttle and $7\ ^1/_2$ four-stroke horsepower mean that we fill the 3-gallon tank very few times in a seven-month season, but that's more a convenience than an economy.

To appreciate a boat like *Puxe,* you have to picture using it. At cruising speed we speak to each other with the same voice volume that we use in the livingroom. Engine noise is slight, and water noise is virtually nothing. We are not ashamed to look behind us and see our small wake lapping the bank. Animals are little disturbed, and plants and their roots not at all. There is time to notice what we're passing and discuss it, but even with the tidal current against us we get past the landscape smartly, about at the speed you'd make on a relaxed bicycle ride. The steering reacts like that on a sailboat, not like a skitterish planing powerboat. "Low-pressure" is what Farmer calls such boats, and what he means is that there isn't strain on the boat itself—or on skipper and crew. It's not like a roller-coaster ride or a gladiatorial combat between automobiles. At the same 10 knots in a runabout, you'd have to stand up to see over the bow. But we're relaxed, enjoying the ride and the scenery and each other.

# Buy-boat

The grace and ease of a spin in *Puxe* may remind some of us of a horseback ride, but the Buy-boat is a horse of another color. Philip Leasure was soon retiring from teaching at a New England community college and was casting about for ways to fill his time. Perhaps he could trailer a boat to Florida and assist in the recovery of a Spanish galleon or two. For that he would need a boat with a hold.

From what I've heard about Mel Fisher and his sunken treasure operations, I doubt that he would entrust a cargo of doubloons to any retired professor. However, a client is not looking for that kind of advice from a designer. We must fall in with the client's program and limit our advice to how it can best be carried out. Two years ago Phil Bolger and Friends went very far with the designs for a self-sustaining liveaboard houseboat for the Arctic Ocean. That one didn't get built, and to date the treasure-hunting Buy-boat hasn't been built either.

The real Buy-boat problem was not the hold but the enclosed head. Many clients make an enclosed head the first priority of design, and certainly it can be achieved in a boat this size. But what it always means is that somewhere there is a woman whom the client would like to take along on his voyages and who has no intention of going. If the private toilet room is drawn, something else does not suit. If the boat is designed, it isn't built. If it's built, it isn't used. The problem is always ascribed to that particular boat, not to boating in general. What a relief it is to remember the straightforward Guideras. Gloria said she didn't want to sail anymore, and John said, Okay, I'll get me a Melonseed and sail by myself.

Leasure was attracted by *Elegant Slider,* a 21-foot foam-sandwich diesel cruiser that was on the cover of my first boat book. The lines were taken from a 26-foot Elco of the 1920s, scaled down because foam sandwich weighs so much less than traditional carvel construction and takes up less of the boat's interior volume. There wasn't room for a private head in her, but Leasure thought that by bringing her up to maximum trailering size (*Slider* was only 7 $^1/_2$ feet wide) some other amenities could be worked into a boat that would cruise at hull speed or better with very small power.

For the preservation of many a set of Elco lines we are indebted to Weston Farmer. He was a company draftsman, and providentially he took home quite a few blueprints of their drawings. When Elco's yacht factory burnt in 1948, the company was busy building atomic submarines at another site, and years later Farmer redrew the plans and barefacedly sold them. The 26 has round bilges and rockers up aft until virtually no transom is immersed. Waterline length-to-beam is only 3.3:1, but under way it is undoubtedly more because the wave train of a hull creates a hollow amidships where the boat is widest, and the round bilges slope sharply inward from waterline down. Is this truly a long, narrow boat?

In final form the Buy-boat I drew for Leasure was 4.2:1 on the water, with the same round bilges, and might be over 5:1 under way. Leasure wanted to put only 15 horsepower into her, which is less engine than one would expect to find in a 30-foot sailboat nowadays. *Slider* had 1 horsepower for every 182 pounds displacement. With 15 horsepower, every horse would be pushing twice that much weight in the Buy-boat. I specified 15 to 30 horsepower for her, and hoped Leasure would think better of his choice later.

The boat was meant to be attractive looking and pleasant to use. Leasure wanted standing headroom in the wheelhouse-saloon, but it wasn't required in the bunk cabin and head. The two cabins were to be separated by some deck. Ordinary powerboat practice would put the wheelhouse forward and the bunks in an aft cabin, but I was thinking of the majesty of the old Chesapeake buy-boats, some of which can still be seen, although they are now reconfigured as yachts. In the days when oystering regulations were stricter and only skipjacks were allowed to dredge most days of the week, buy-boats would carry the oysters from skipjack to market. They were bigger than the skipjacks—some may have been 100 feet—but were similar in conception, with a crudely shaped V-bottom and enormous scantlings. To build and maintain them, wood had to be cheap.

The traditional buy-boat wheelhouse was several steps up from the deck, and of course that wasn't possible in a boat 30 feet by 8 $^1/_2$ feet. It was also very short compared to overall length, to leave a big deck for oysters. An early version of the Leasure Buy-boat did have a significantly shorter house, but it had to be lengthened when Leasure decided that the dinette needed to break down to make a bunk for guests. You must judge for yourself whether the aura of the workboat has been saved in this tiny replica.

What really saves the design is not having the private head in the saloon, where the standing headroom is. No matter where it goes, it blocks a quadrant of the helmsman's vision until a boat is big enough to have a wheelhouse several steps up from the saloon. The bunk cabin in this boat is about what you'd expect in a 20-foot sailboat, and although it has sitting headroom it wouldn't suit me. On most cruises there are some days when boats stay in port, and I'm likely to spend most of such a day reading in the bunk. In the Buy-boat I'd head at once for the saloon with its big portlights and wouldn't enjoy the periodic returns to the forward cabin to answer nature's calls.

What I like least about Buy-boat is the sailing rig, which the real buy-boats never had. When the boat was being drawn Leasure owned a Peep Hen, so presumably he sailed, although that boat couldn't have taught him anything about sailing well. He thought that 25 square feet of sail would be a "get-home rig," but I doubt it would even steady a 5,900-pound boat. It's just expense and clutter. Modern diesels are proof against almost anything but fuel contamination, and if that should happen, modern radios summon help anywhere the boat is likely to find itself.

The Buy-boat has two wheels but only one set of engine controls. In fair weather and uncongested water, the right-handed door would latch open, and if the skipper were steering from on deck, controls and instruments would be only a step away. The side deck is 22 inches above the cockpit sole, too high for comfortable sitting and too low for leaning, but a folding chair would make the cockpit helm a nice place to enjoy the breeze.

Two wheels operating off one hub is not a usual arrangement, and such hardware might be difficult to find in steering equipment catalogs. The idea was to use

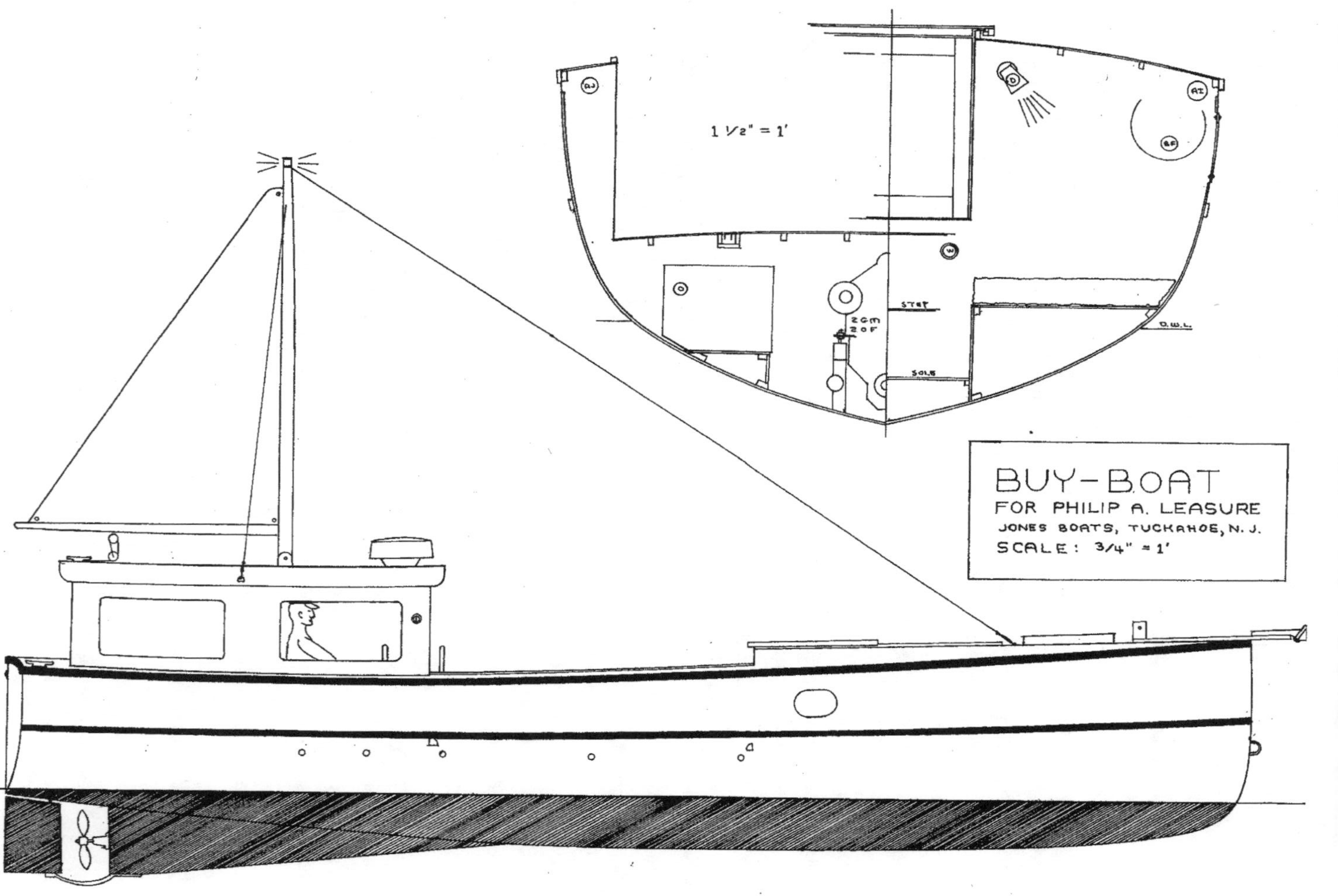


Buy-boat plans (a)

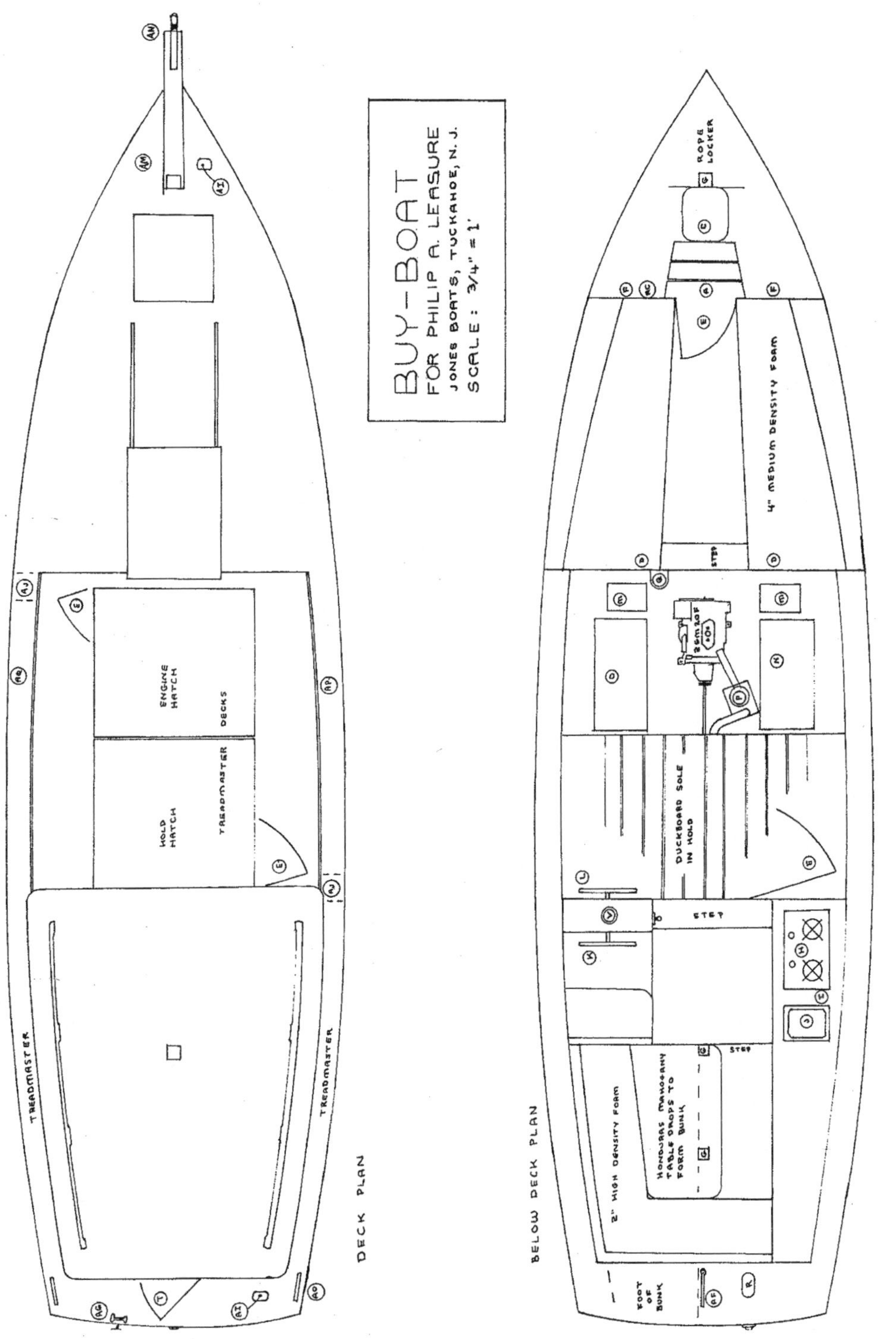

Buy-boat plans (b)

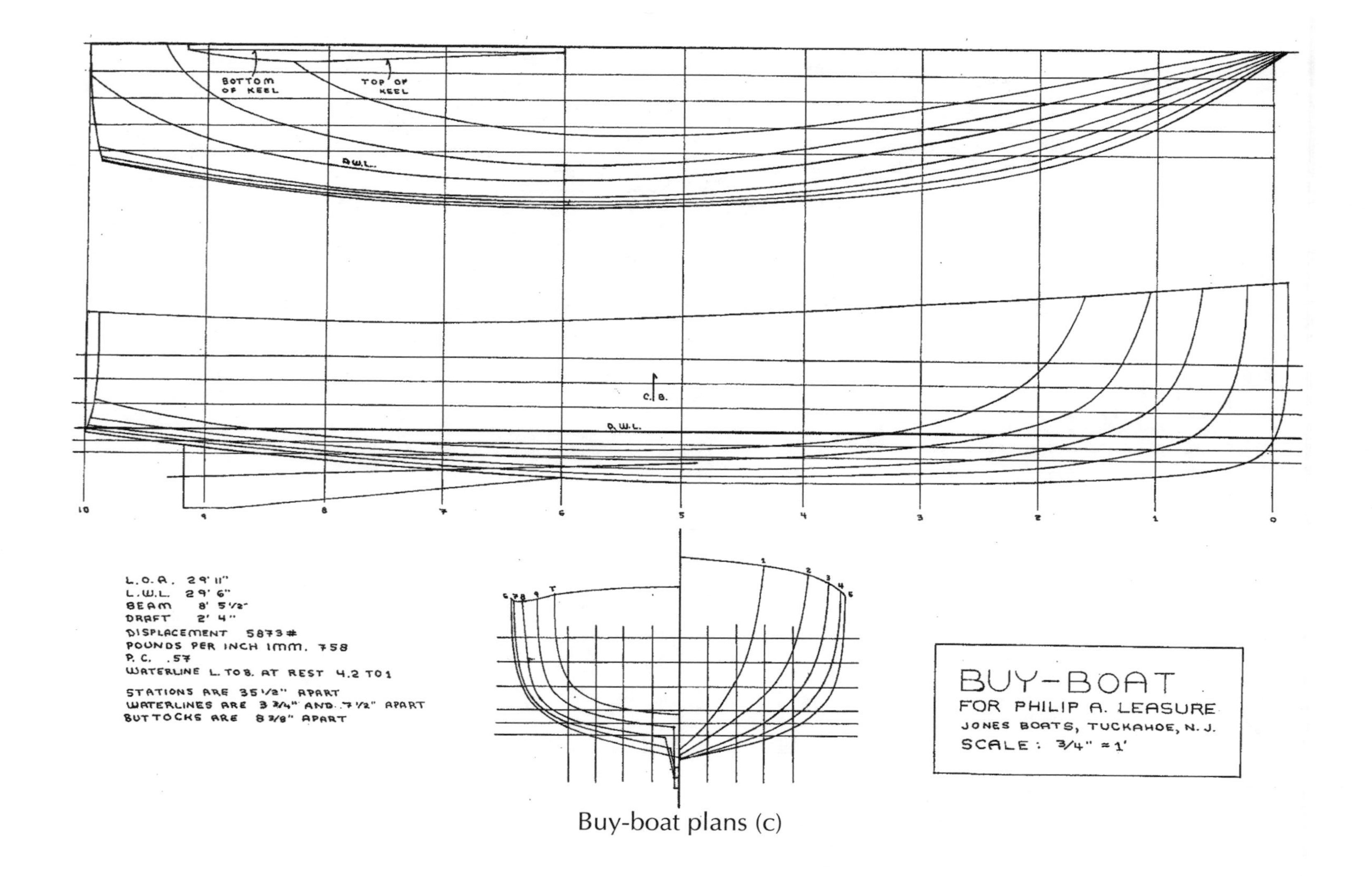


Buy-boat plans (c)

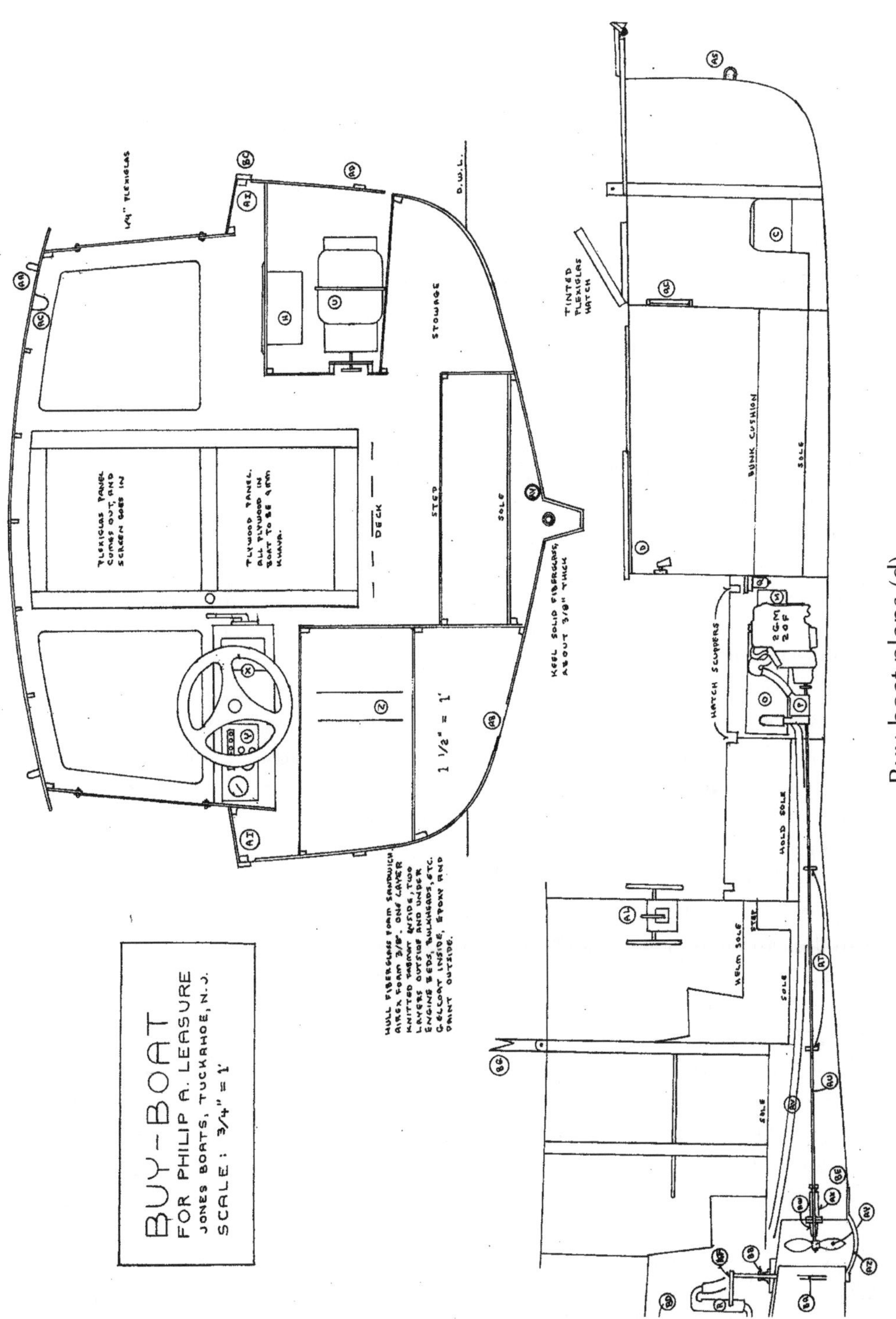

Buy-boat plans (d)

a 10-inch bronze shaft and a drum with cables. Getting the cables through the bilge would have required a number of blocks, but it would have been inexpensive compared to steering systems from a catalog, and reliable, too. We thought the Buy-boat could be built in my shop for less than $50,000, but that was ten years ago. That price would not have included electronics.

Even with the helmsman sitting in a chair by the wheelhouse door, the noisy diesel is some distance away, and in the wheelhouse it should barely be audible, with two insulated bulkheads between. It's amazing how often the wheelhouse of a powerboat is sited directly above the diesel, and it's also amazing how popular these rackety engines are. They have driven gasoline marine engines from the market, except in speedboats where light weight is paramount.

Diesels are said to be safer, and every year a few gasoline inboards do blow up. It usually happens after fueling, when the eager skipper has not ventilated the bilges for the prescribed five minutes before pushing the starter. With blower ventilation, gasoline engines are safe; and with modern ignition and no breaker points, they are reliable. They are infinitely smoother and quieter; yet how often we see an older powerboat converted from gasoline to diesel. The next year a flying bridge will spring up from the cabintop, as the skipper scrambles to escape the noise and vibration of his new engine!

The Buy-boat hold contains only 35 cubic feet, which is not a lot for cargo but generous for deck furniture. Many a bigger boat has no storage space at all. Hold and engine hatches are scuppered so that the entire deck is flush, and there's more safety in that than in a diesel engine. Deck area is 54 square feet, about as much as on our catamaran *Dandy,* so the Buy-boat could host a good party.

Dave Kelly from Arizona bought a set of plans not long ago. He thought to build her in 1-inch Douglas fir strip planks, probably because he was used to building that way, but it would take some time in a hull this size. I used to think that strip planking made the best imaginable wooden hull with no voids or corners to trap moisture and dirt. More recently I have seen older strip-planked boats that were big enough to have bulkheads, like the Buy-boat. The strips can shrink, and as the plywood bulkheads don't, the strips come apart. The splits or parted seams with their frequent nails are not easy to make tight again.

In addition, 1-inch fir weighs 3 pounds per square foot, and the foam-sandwich layup specified was expected to weigh half that. Dave had some other ideas that weren't going to make the boat lighter and some very optimistic expectations of how fast a 35-horsepower engine would move her. However, he had been through a mail-order yacht-design course, and he eventually concluded that the Buy-boat wasn't his ticket from high, dry Sierra Vista to the promised land of water.

As no example has yet been built, the performance of this boat can only be conjectured. However, before we sold *Slider,* Carol and I motored her from New Jersey to Key West and back again to central Florida. She had a Universal diesel delivering 18 horsepower at 3600 revolutions per minute, and I had such hopes for her performance that I gave her a 13-inch-by-10-inch propeller. The diameter comes from a nomograph in *Skene's* that is entered with shaft horsepower and rpm, and you'd have to be a moron to get that wrong. Pitch is harder to figure, and wishful thinking can play a part. Figuring prop slip at 25 percent, 3600 rpm would give her 11 knots. Even if the engine would only rev to 90 percent of its rating (as they often do under load) we'd still have 10 knots, or a speed-to-length ratio of 2.2.

With that prop the engine wouldn't rev to 3000 rpm, even downwind. The load was too great. Chastised, I had an inch of pitch taken out of the prop before we started for Florida, and that did allow her to rev up fully. In Florida, in the tideless Intracoastal Waterway on a windless day, we timed the boat between bridges. At 2400 rpm, which on the way down had seemed the best compromise of speed, noise, and vibration, she did 6 $^1/_4$ knots, or just over hull speed. At 2800 rpm she did 6 $^1/_2$ knots, or 4 percent more speed for 17 percent more revolutions. According to the Universal literature, horsepower and fuel consumption had also increased about 17 percent, and the racket she made and the shaking she did increased out of all proportion. So too did the wake. Like Emil's overpowered flatiron, *Slider* was digging herself a hole in the water.

The Buy-boat would be faster because she is longer: hull speed would be more than a knot greater. But I don't think her lines, however graceful, would lift her much beyond hull speed. She's 25 percent narrower on the water, and that certainly couldn't hurt, but she would be pushing a lot of pounds with each horsepower and she has proportionately less lifting surface aft. These boats with good lines move along at hull speed with very little fuss, but in my opinion they're a far cry from genuine long, narrow hulls.

The last time I heard from Phil Leasure he was planning to buy a production 23-foot fiberglass cruiser of the kind where everything folds up and becomes something else. He still has the same girlfriend, so I hope that the head at least doesn't fold up.

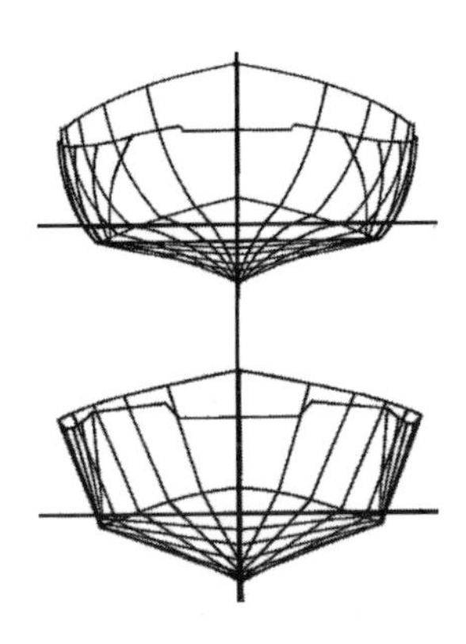

# 5

# *Sailing Multihulls*

This chapter discusses three catamarans and one trimaran that were not included in my last book, *Multihull Voyaging*. Indeed, three of them weren't drawn when it was written. A single chapter cannot describe the "multihull movement," as it used to be called, in as much detail as a whole book. But a few things need to be said about the advantages and disadvantages of cats and tris in general before getting into the designs themselves.

The greatest advantage of a multihull cruising boat is not speed (although a good one will clip the wings of monohulls that are bigger and more costly) but the stable platform. Bouncing around and heeling over in a daysailer is entertaining, and the inherent complexity of multiple hulls is a poor choice for a small boat; but by the time your boat is big enough to sleep and cook on, by the time you think of taking a cruise and living aboard for days or weeks or years, the sane skipper chooses to do it upright, not heeled or rolling. In *The Hungry Ocean,* fishboat captain Linda Greenlaw, describing her last breakfast ashore with the boat's owner, remarks that the coffee mug in the diner does not skitter across the table, "something I would not see for the next thirty days." Her boat was a diesel-powered steel 100-footer. God help the folks out there in 30-foot monohull sailboats!

Another advantage of cruising multihulls is that they don't have ballast. In a good daysailer crew weight is ballast enough, but in a boat big enough to have accommodation crew weight is a very small part of total weight (except for racing, when every ounce counts, so the crew has the pleasure of hanging out over the rail upwind and sitting on the keel downwind). Monohull designers brag about how high a percentage of total weight the ballast is, but friends, no ballast at all is every way better. The square root of sail area can be compared to the cube root of weight by various formulas, and they yield a very good indication of speed. Much ballast is not a virtue but a defect in cruising boat design, because the desired stability—undreamed of stability, in fact—can be achieved by simply having more than one hull. And think of what happens to a ballasted boat when she's holed or turned over: blub, blub, blub!

My multihulls are simple and sturdy, but weight has been watched closely. Two or three narrow hulls don't carry weight as well as one fat one, and whatever extra the boat weighs must be subtracted from payload. Gear should also be simple and light. Propane is wonderful for cooking, but other systems, such as plumbing, electricity, and especially refrigeration, are the main sources of trouble on any boat and too much weight for multihulls under 30 feet. Dry-cell running lights, kerosene lanterns, water in gallon bottles, and a bucket head or porta-potti are the best choices. Because multihulls are not deep in the water, having 6-foot headroom makes too much windage in the smaller ones and in any event contributes very little to comfort except when pulling up your pants, which can usually be done in an open hatch. Comforts that matter more are good seating with backrests, bunks at least 6 feet, 8 inches by 2 feet, 6 inches, and dry cabins with cross ventilation.

These were our opinions when Carol and I started cruising on multihulls, and twenty-five years' experience and 50,000 miles of sailing (including six transatlantic passages) have only reinforced them. Other people prefer to "play house" on a boat. Every device of a house on land is there, but miniaturized for cuteness. "This is the bathroom, but we call it a head. That looks like a shower stall, but it's a sit-up bathtub, too. We have a little trash compactor under there: it's really not so hard to get at. Then there's a private stateroom here, right up in the bow, with a dressing room *en suite,* and back there. . ." Some multihulls try to cater to these folks, but the ones in this chapter do not.

One thing I've learned since *Multihull Voyaging* was published is the extra speed and fun of using an asymmetrical spinnaker. A symmetrical balloon spinnaker can do without a pole on a multihull because it can be tacked to the windward hull, but it is essentially a sail for pushing the boat downwind, and often a multihull covers ground faster by "tacking" downwind (that is the term always used, although actually you are jibing downwind) in a zigzag pattern.

Some monohulls can surf downwind or on a reach, exceeding hull speed in brief bursts; and very light ones can sometimes plane downwind thanks to Uffa Fox, who figured out the hull form that makes it possible. But the narrow hulls of a multihull allow it to cut through the water without rising up, just as long, narrow powerboats can do. Our *Dandy* has a theoretical hull speed of 6.4 knots, but reaching down Long Island Sound a few summers ago, with the tidal current against us most of the way, we covered 20 nautical miles from one breakwater to another in 2 $^{1}/_{2}$ hours. *Dandy* is a fairly staid design, and no doubt some other multihulls could have done it faster, but we were never planing or surfing. We were just going through the water, sailing with main and jib, not the spinnaker.

For a comparison of the two spinnaker types on a multihull, imagine a 10-mile course dead downwind in 10 knots of wind. Setting a balloon spinnaker along with the main and steering a direct course, you could cover the distance in perhaps 2 hours. Apart from watching the compass and tweaking the tiller occasionally (to keep the main from jibing!), there wouldn't be much to do. The 5-knot apparent wind would hardly dry your sweat. If you set an asymmetrical spinnaker and headed up 45 degrees with the intention of jibing halfway along, a good multihull is likely to accelerate to 8 knots or better. That's because the apparent wind is now just forward of the beam and blowing at better than 15 knots. The distance to be covered is 14 miles, but you arrive sooner and your sweat soon dries because the sails are now foiling, not just pushing. Best of all, the sailing is exhilarating.

Carol and I first studied asymmetrical spinnakers when we began racing *Dandy* at handicapped regattas in New England. Some tyros overdid it, especially in light air, and we watched them sheet in hard on both main and spinnaker, zigzag busily back and forth across the course, and make negligible progress toward the next buoy. Heading up more than 45 degrees from the direct downwind course seldom profits. When the wind is strong enough, the fastest and most pleasant route is to head straight for the destination. When they get going fast enough, all boats throw some spray around, and even multihulls can be made to bounce on waves, with strong wind and skipper encouragement. Going straight downwind under spinnaker is a kind of first reef, and one hopes that's all the reefing that will be necessary.

Various systems can be used to jibe a spinnaker faster, and they do work. Some multihulls use a bowsprit, which, with a system of ropes and pulleys, can swing the spinnaker tack to port or starboard. Perhaps with two sheets like immensely long jibsheets one could jibe the spinnaker without leaving the cockpit. Catamarans are sometimes seen with an extra rope or wire bridle between the bows. The tack is attached to a block on the bridle and can be jibed with an arrangement of lines and blocks. The weight, expense, and clutter of this gear justifies itself in racing but not in cruising. However, we do set our asymmetrical spinnaker on every occasion that we can, and we find that it makes sailing downwind as challenging and as much fun as sailing upwind.

In *Multihull Voyaging* I urged skippers to race in handicap fleets, and for four seasons we did so ourselves and might still be doing so if we lived nearer the races. However, whereas in one-design racing you are racing against other crews on an equal footing, in handicap racing you are racing against your handicap. Are you better and is your boat better than the handicapper thinks? Then you win. Multihulls differ from each other more than monohulls do, so handicapping them is harder, and because the fleets are newer and smaller and less well organized, the handicapping is more haphazard. So you may do well or badly, but you are not likely to get much satisfaction from it. After four seasons, we were no longer willing to schedule our annual summer cruise to New England to coincide with the regatta schedule.

In addition, most multihull events are just one class in larger regattas, so multihull sailors (most of whom are decent enough) are thrown up against the truly loathsome creatures who race monohulls.

Let's talk about more pleasant subjects. You may have looked ahead and noticed that none of the four boats in this chapter has a cockpit. Looking through C. P. Kunhardt's *Small Yachts,* first published in 1885, it appears that every one has a cockpit; but we have discarded a great many yachting fashions from 1885, and perhaps cockpits should have been one of them. Chapelle shows us that small workboats, in which hull space could make the difference between profit and loss, often lacked cockpits. Isn't our space precious too? Tom Colvin, who was designing impressively unorthodox monohull sailboats thirty years ago, many of them steel and junk rigged, could wax apoplectic about cockpits: they stole space, they weakened the hull, they were slow to free after pooping. That's all true, but some people still want cockpits.

Under way a cockpit gives a sense of security because you are surrounded by seat backs or at least coamings. But in that rogue-wave situation, if there is such a thing (and I've never seen one), a cockpit is not secure because there's nothing to hang on to except the tiller or wheel, which often breaks. Many people lost at sea from sailing yachts fall or are thrown out of cockpits. If you are sitting in a companionway you can hang on with both hands and perhaps curl your legs up under the bulkhead; but the greatest real security comes from a harness or perhaps the bitter end of the mainsheet tied around your waist.

It is true that under way in fair weather a half-dozen people can gather in a cockpit and there converse, much as if sitting around a table. On bigger boats the cockpit sometimes does have a table. But on smaller boats too many people in the cockpit throws the boat out of trim, and on a light multihull the effect is pronounced. Besides, imagine a cockpit aft of the connecting beam on any of these multihulls. At its forward end there wouldn't be room for two people to sit facing each other, and getting them over the beam might require a ladder. Put a cockpit forward of the beam, and little is left of the accommodation.

In port people like a cockpit because they rig an awning and sit out there, like ducks in a dishpan. Perhaps in a monohull that's the only possibility, but on a catamaran there's that big wonderful bridgedeck that can also have an awning and makes a nicer livingroom than any structure that could be put on it. It catches the breeze on anchor, while in a cockpit the breeze is blocked by the cabin. Even on a trimaran like *Night Heron,* where light weight and folding arrangements dictate nets rather than rigid decks, the nets are infinitely nicer places to relax and study the harbor than any cockpit could be.

For leeway prevention the Weekender has fixed keels of less than 1:1 aspect ratio. They aren't as efficient as a daggerboard would be, but they do allow some semblance of accommodation in this tiny boat. The other designs have daggerboards, set to one side and canted about 10 degrees, again in the interest of uncluttering the accommodation. That looks wrong to some people who want the daggerboard on centerline and upright. "Symmetry is always in fashion," says Carol. But monohull sailors are content to sail to windward at 15 or 20 degrees of heel, with daggerboard or keel at that angle. The only efficiency lost is that the board isn't quite as deep in the water as it would be upright. At 10 degrees that loss of depth is 1 $^{1}/_{2}$ percent and well worth it even on a sporty cruising boat.

Three of these four designs are trailerable. In *Multihull Voyaging* I didn't have much good to say for trailering, and I still don't like it, but many people still desperately want to do it. Thirty years ago people were wanting to trailer monohull sailboats, and mass-produced ones sold like hot cakes. As these boats were plastic and not in the water enough to suffer osmosis, you would think that some would still be with us; but you never see one at a launching ramp. The few trailersailers that you still do see are at docks or on moorings, just like the sailboats whose designs were not compromised for the sake of trailering. Presumably, the rest of the immense trailersailer fleet are in backyards.

The skippers who often use launching ramps often have 15-foot outboard runabouts that tow behind a car with little fuss and can be launched and ready for use in a minute or two. There are no hulls to arrange, no rig to erect. Probably in twenty years the multihull trailering mania will have gone the way of the monohull one. In the meanwhile, three of these boats offer solutions to the problems, while *Dandy II* was designed purely for the water.

# Weekender

The English geographer Richard Hakluyt, writing in the late sixteenth and early seventeenth centuries, tells us that in 1583 Sir Humphrey Gilbert was returning from America to England in a 10-ton frigate, followed by a larger vessel. Off the Azores a gale arose, and in one wave train (big waves usually come in threes) the frigate was nearly lost but did recover. Sir Humphrey, reading a book, called over to the larger ship, "We are as near to heaven by sea as by land." Lamps were lit, one of them presumably so that Sir Humphrey could continue reading, and at midnight the lamps suddenly disappeared, as the frigate was "devoured and swallowed up of the sea."

For most of us who go to sea these days, heaven is not our destination, and we would like to know what kind of vessel will keep us from being "swallowed up." The Weekender is hardly the best choice, although she is fully decked and the frigate probably was not, so she would not fill with water in an instant. In a Wharram Tiki 21 Rory McDougal sailed from England to Australia and home again. Let us compare some numbers for the two catamarans:

| | Weekender | Tiki 21 |
|---|---|---|
| waterline length | 16 ft, 8 in | 18 ft, 6 in |
| weight | 750 lbs | 650 lbs |
| interior volume | 128 cu ft | 113 cu ft |
| sail area | 150 sq ft | 208 sq ft |
| hull centers | 6 ft, 4 in | 9 ft |

A standard rule of thumb for catamaran design is that the centerlines of the hulls should be half the waterline length apart. Although big heavy cats can get away with less distance (wind speed and wave height do not increase with boat size), small, light ones should follow the rule pretty strictly because it is the most

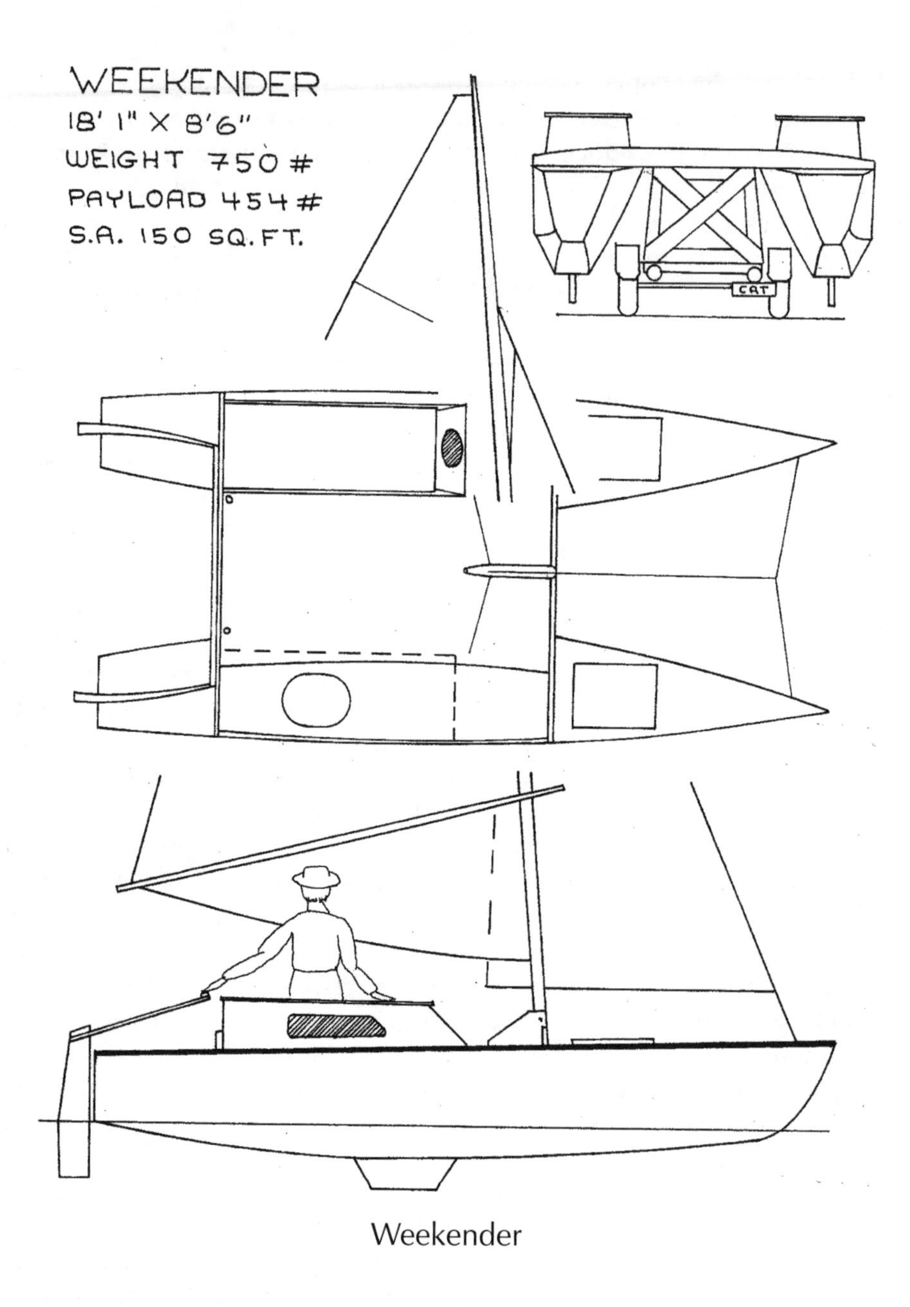


Weekender

important number for stability. It might seem that the greater sail area of the Tiki would even things out, but sail area can be reduced, and presumably Sir Humphrey's frigate wasn't under full sail either.

Jim Wharram says that his Tiki 21 is trailerable. The hulls are lashed to the beams, and each is supposed to weigh only 200 pounds, although Jim's weight calculations are notoriously optimistic. At any rate, the two hulls, the three beams, the deck and netting, the rudders and tiller bar, the mast and boom, and any personal gear are put on a trailer less than 8 $^{1}/_{2}$ feet wide, and off you go. The idea with Weekender was simply to float her onto a trailer, drop the mast, and go.

There are several other designs on the market for cruising catamarans under 8 $^{1}/_{2}$ feet wide, but naturally I think Weekender the best. Others tend to be heavier, and the additional weight requires fatter or deeper hulls to carry the load and a bigger rig to drive it. If the hulls are fatter, they must be closer to each other, and a length-to-beam ratio of 9:1 is needed if hull speed is to be readily exceeded under sail. In a catamaran with flared hulls (and that includes all my designs and all of Wharram's), the waterline had better be even narrower, because by the time exceeding hull speed is a possibility the boat is usually heeled several degrees, which is enough to push the leeward hull down several inches. By contrast a trimaran like *Night Heron* can have an 8:1 flared main hull because heeling her takes weight off that hull and raises it several inches.

Weekender won a design contest sponsored by a nautical publisher. Phil Bolger was the judge—the contest ran in the days before we got to know each other well—and he thought she would be "a handy boat that would be exciting to sail without being very demanding." Cliff Wade isn't so sure. He built one for sailing on the artificial lakes in western Missouri that are supposed to control flooding on the Missouri and Mississippi rivers. With his stepson aboard, he launched her in 18- to 23-mile-per-hour winds and "had a blast." He carried the full main and no jib, which a lot of people find an easy way to reef, and apparently he was able to control the boat. My idea was to carry the jib with a single reef in the main and drop it when the second reef in the main was tucked in, because sail balance would then be the same as with plain sail.

Cliff tried her again another day, all alone and with winds gusting to 30 miles per hour. "It was a fairly hair-raising experience," he said, and that isn't surprising. Suppose that he weighs 200 pounds and was sitting on the windward cabin and that he again used full mainsail only: theoretical capsize should have occurred at a wind speed of 28 $^{1}/_{2}$ miles per hour, with wind on the beam and the sail sheeted flat. "I believe when I go by myself from now on, I'll sandbag the bilges and pay more attention before I pop out of a cove," he said. I don't know what happened after that, but he did buy plans for a Brine Shrimp. Weekender has two and a half times the stability of a Hobie 16, but if the Hobie capsizes she can be righted by her crew.

It seems unlikely that on those lakes Cliff ever used the overnighting capability of his boat, and it seems too bad to put up with the clutter of cabins on a boat used only for daysailing, but a great many people do it. Weekender's accommodations aren't miserable: the bunk boards are 8 feet by 2 feet, and the mattress is a bit higher and therefore wider. With the mattress removed (deflation would be the ideal method), sitting headroom is 2 feet, 9 inches, which allows you to sit down if not up. The bunk boards are glued in, but the oval panels lift out to allow your feet to be 12 inches lower than your seat and for stowage of spare clothes, a Sterno stove, and a few other things that you might bring on a weekend voyage. On our first cruising catamaran, Carol and I slept on a bunk very little wider than these, but twenty-five years later I would come over for a goodnight kiss before bedding down in the other hull.

On deck the space is also limited but possible. The bridgedeck is a single 4-foot-by-8-foot sheet of $^{3}/_{8}$ -inch ply, so most often the most comfortable place to sit

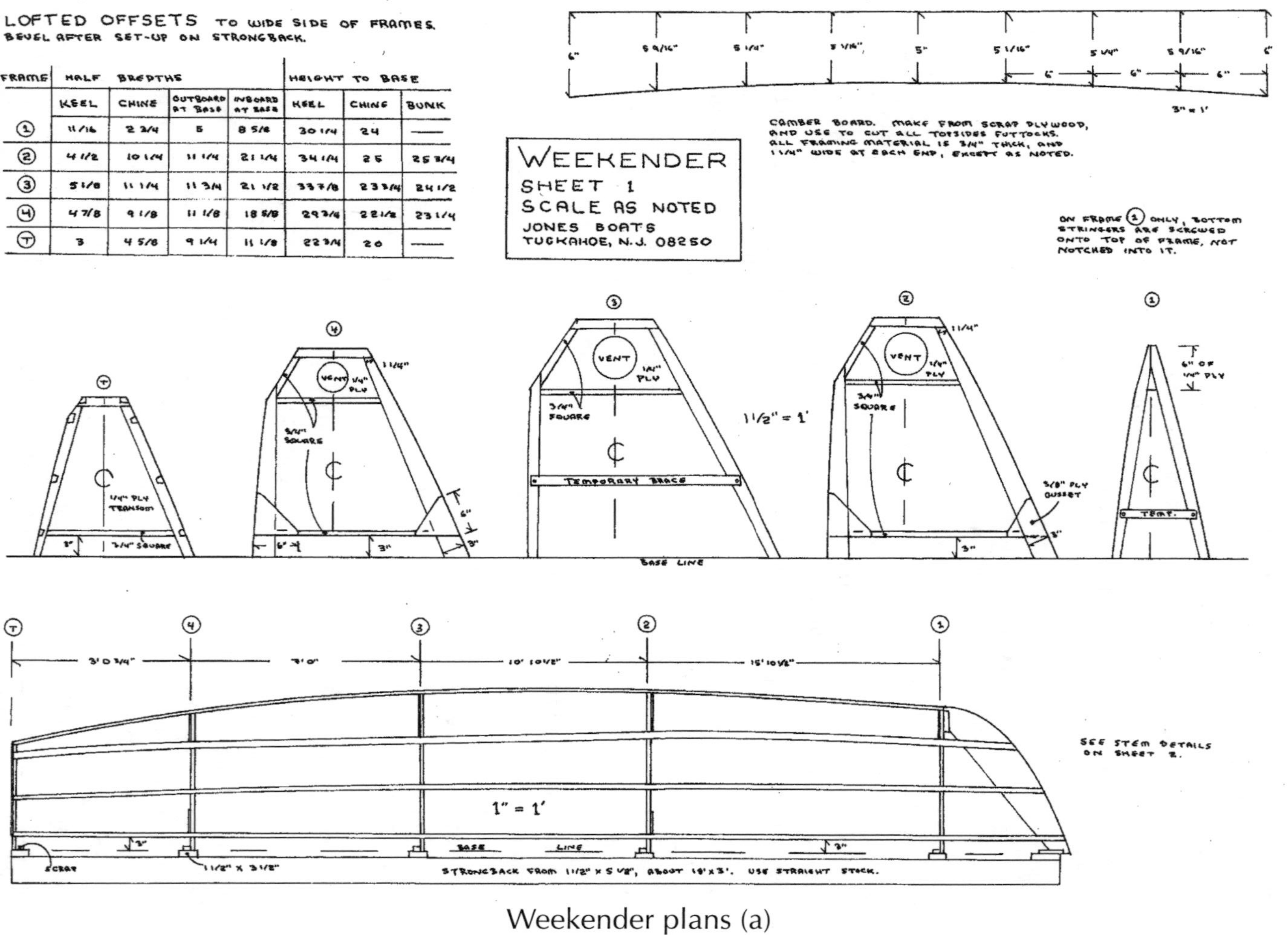


Weekender plans (a)

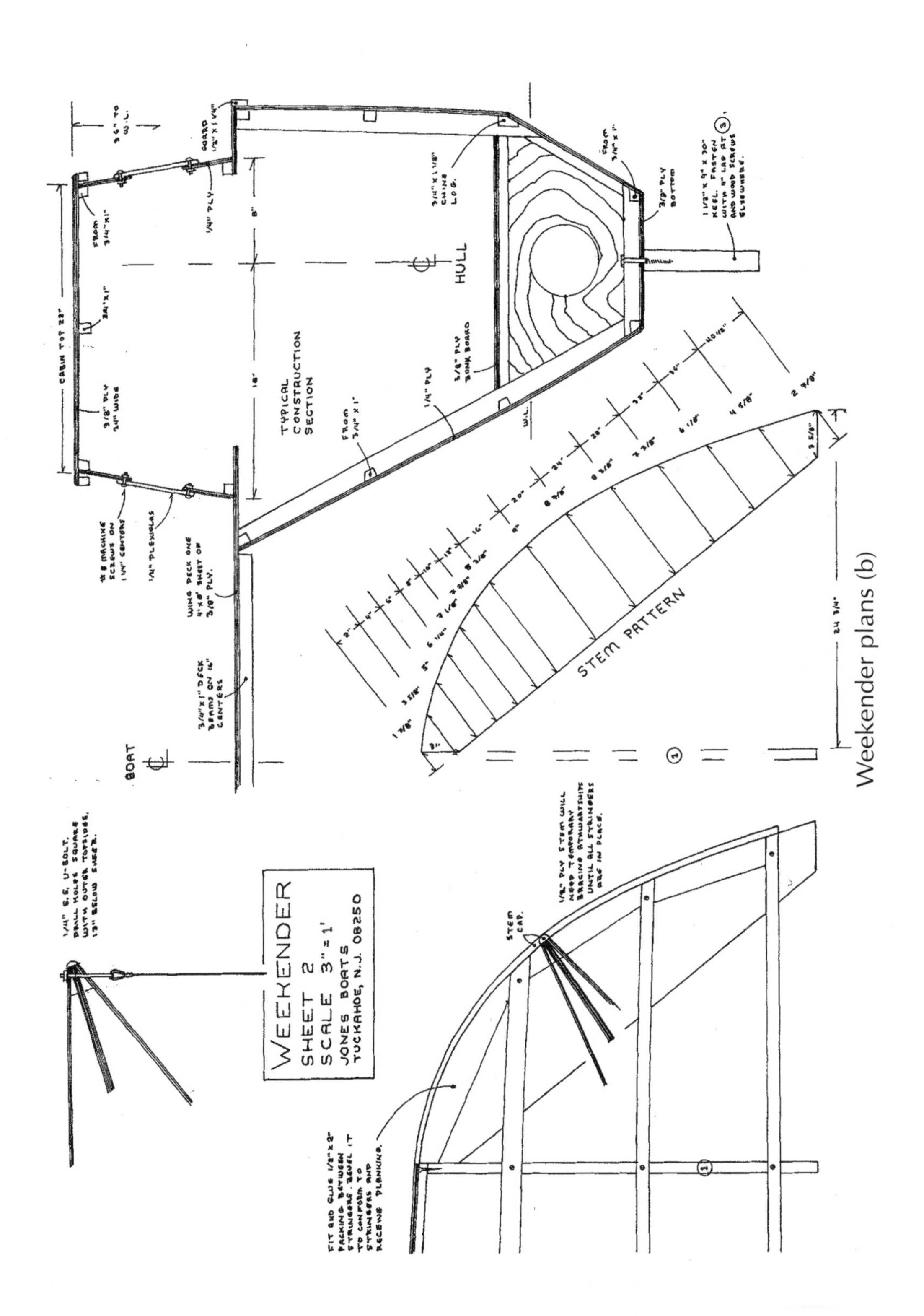


Weekender plans (b)

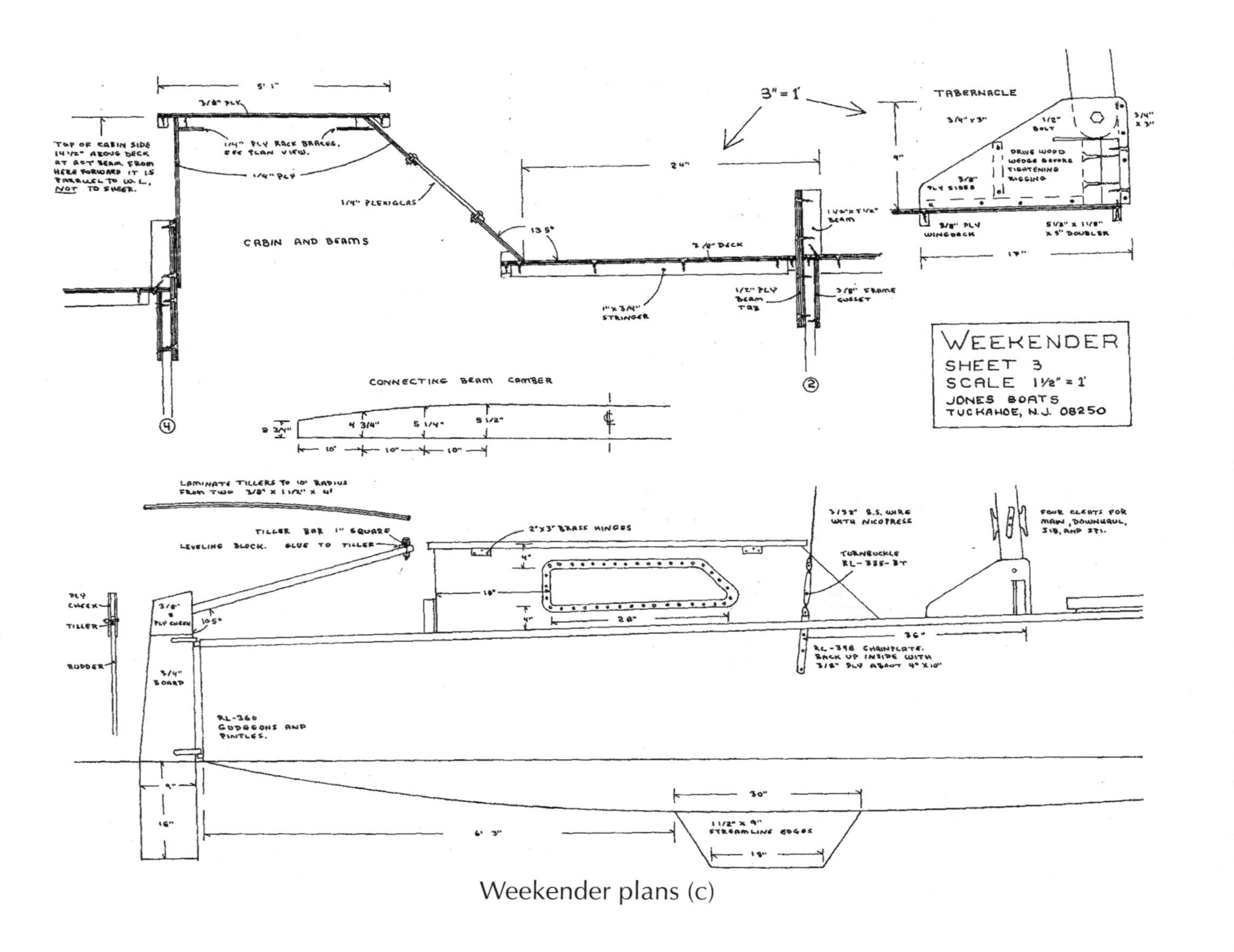


Weekender plans (c)

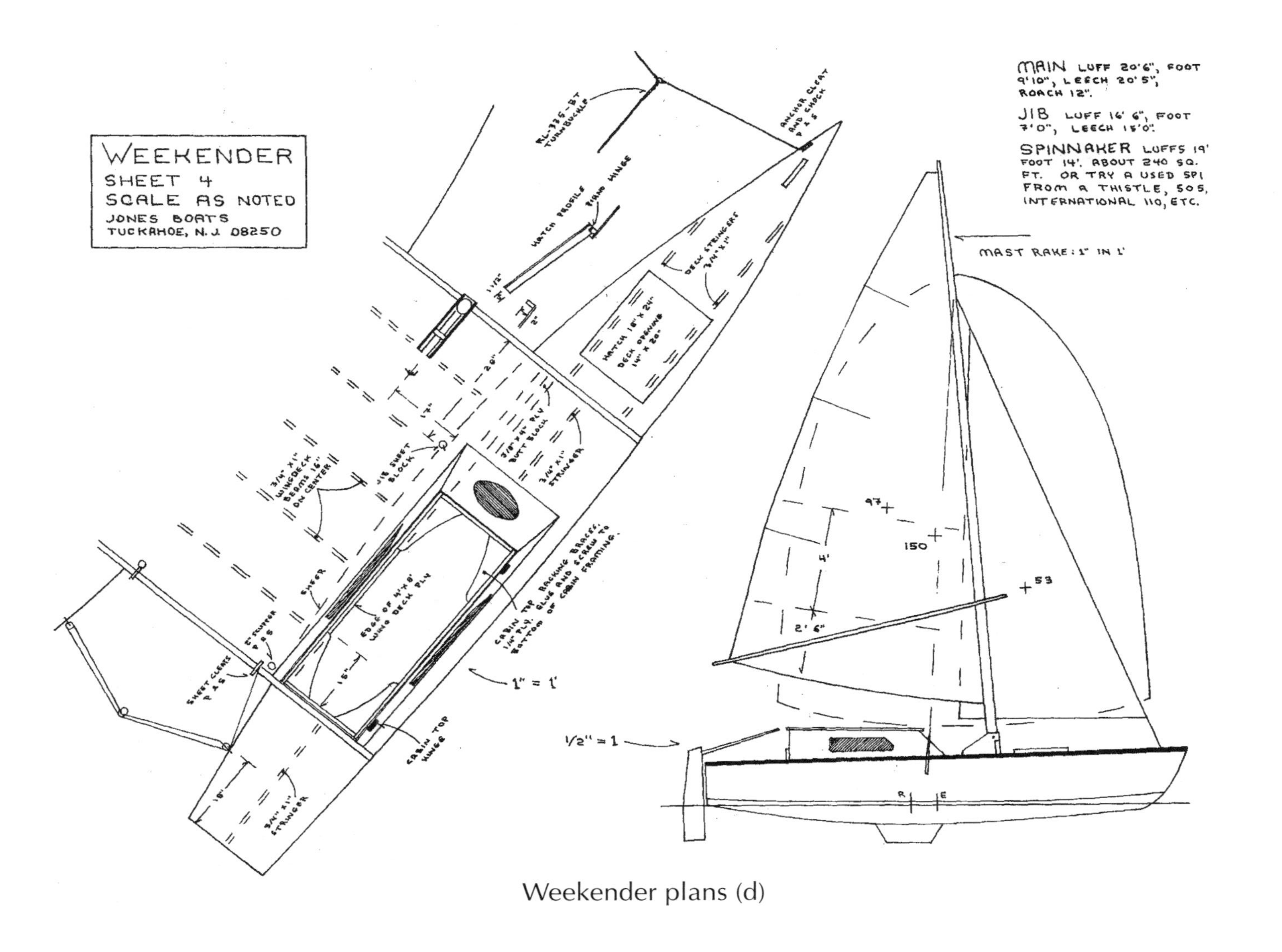


Weekender plans (d)

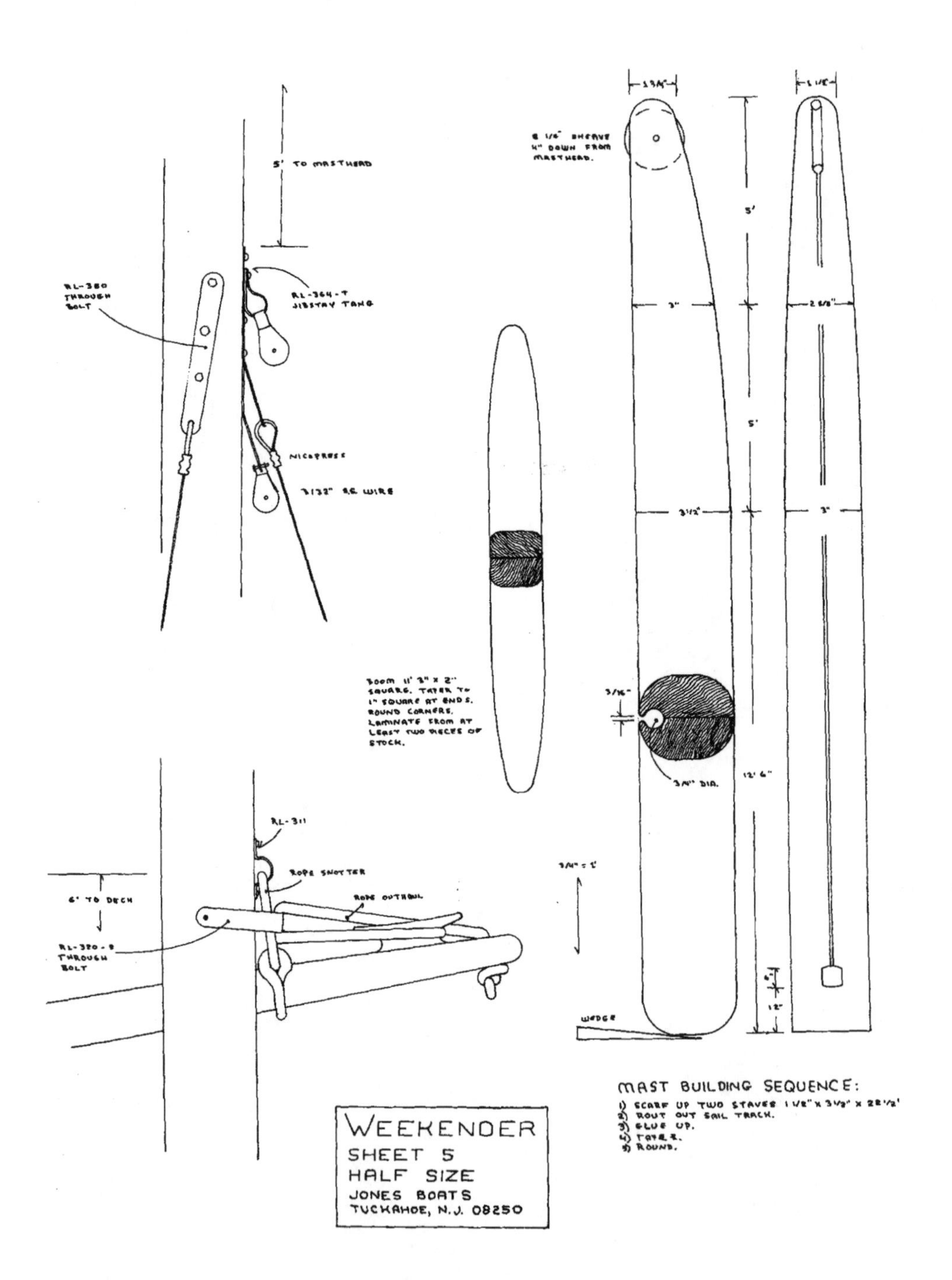


Weekender plans (e)

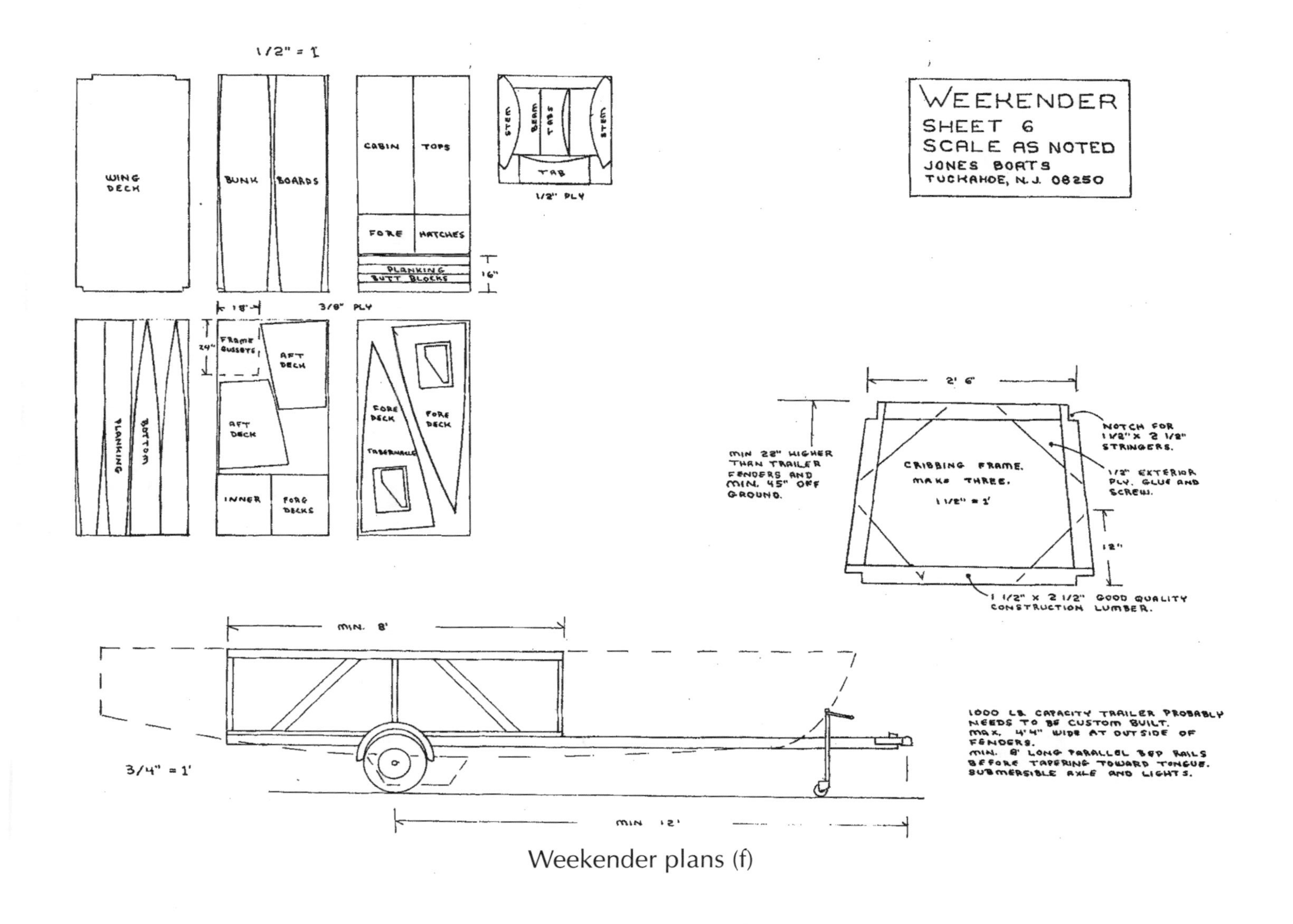


Weekender plans (f)

is on a cabintop, although that does not provide a backrest for lounging in port. The cabintops themselves are hatches with hinges outboard and can be swung up and tied or clipped to a shroud. They don't provide cross ventilation, but they are 4 feet long, so some air should find its way below.

With only a 454-pound payload, gear has to be selected very carefully (on any multihull it's a good idea to think, before going ashore, what piece of gear can be taken ashore and left there). If two people go cruising for a week, their food and drink amounts to 112 pounds (8 pounds times two people times seven days), and it's best to make a stop or two to replenish, rather than set off with all of it aboard from the start. As the boat is best suited to sheltered waters within sight of land, small ports shouldn't be hard to find. For ground tackle, an 8-pound Danforth or a 5-pounder with a bit of chain is enough.

Weekender's keels preclude beaching, and she really is too small as well as too short of payload for any kind of dinghy or outboard motor. Our first two cruising catamarans I rowed, sitting in a forward hatch with a single 9-foot oar between the hulls; but Weekender's narrow hull spacing would allow only a 6-foot oar on the recovery stroke, which would then be nearly vertical on the pulling stroke. A couple of long canoe paddles would move her some, and probably a yuloh would move her better, but I haven't any experience with one.

Like Brine Shrimp and Night Heron, Weekender has a straight sheer, and frames are set up on the strongback so that the sheer is parallel to it, not the waterline. That makes building simpler than does a curved sheer. Not everyone likes the sheer's looks, but it only looks straight when your eyeball is exactly at sheer height, and even then the outward cant of trimaran floats gives the appearance of reverse sheer, as can be seen in the Night Heron drawing. All frames including transoms are set up square with the strongback, which gives the transoms a rake of about 3 degrees.

The mast is drawn to be square with the sheer, so it also rakes 3 degrees, and to adjust the turnbuckles to achieve that, you only need step back from the boat, hold a framing square out at arm's length, and eyeball sheer and mast. However, determining the length of rigging wires is tougher, and in my many battles with Pythagoras I have not always come out a winner. The boat has only five rigging wires—counting the bridle—and a builder may choose to have them Nicopressed or swaged in a shop. Shrimp and *Dandy* have ten wires each, and at that point it is just as inexpensive to buy a 250-foot spool of wire and cutters and a Nicopress tool and fittings. Nicopresses are stronger than the wire they squeeze, and although we have lost several wires over the years (again it's the old problem of stainless steel fatiguing), we've never had a Nicopress fitting give way.

Like all these multihulls, the topsides futtocks of Weekender are curved 1 inch in 4 feet using a camberboard, as shown in the plans. As it does with the bottom of the mini-garvey, this stiffens the panels but does not complicate the planking work. A friend of ours built a 38-foot plywood trimaran whose topsides futtocks curved 1 $^1/_2$ inches in 4 feet, and they were quite difficult to fit.

When installing fixed portlights in a plywood cabin, the cutting and drilling suggestions in the structural Plexiglas section of chapter 6 should be helpful. Boughten plastic framing into which portlights can be clipped tend to trap water

Weekender on a Missouri lake (Photo courtesy Cliff Wade)

and rot plywood. It's best to make a portlight about 1 1/4 inches bigger all around than the cut-out in the plywood, and throughbolt it on about 1 1/4 -inch centers with round-head machine screws. Although Plexiglas does expand and contract at a different rate than ply, I haven't found it necessary to drill oversize bolt holes in 1/4 - inch Plexiglas for mounting on 3/8 -inch cabin sides. That is sometimes necessary with very big portlights and heavier Plexiglas.

It is crucial that the Plexiglas that overlaps the ply be back-painted before installation. If your glass is paper-backed, the portion on the overlaps can be scored and peeled off, leaving the portion masked that you look through. Then the overlap is roughened on the inside and given two coats of finish paint, which acts as a sunblock for the silicon or whatever sealant you use between glass and plywood. No sealant contains ultraviolet shielding, and if paint does not protect it, it crazes and eventually leaks.

The Weekender plans were drawn ten years ago, and if I were drawing them again, the mast is the only component I'd change. It's glued up from two two-by-fours that are first grooved to take the boltrope of the mainsail. Three inches by 3 1/2 inches over all, the solid mast can be well tapered and rounded, and because it has only three wires supporting it, it needs to be stiff. The 40-pound weight may not seem impossible, but a box section like Night Heron's would save pounds aloft, and I'd take the trouble to do it.

Cliff Wade began building in February, presumably after work and on weekends, and he was sailing in October. He cut no corners that I can see, and from the pictures he sent she looks as good close up as she does on the water.

# Brine Shrimp

The original Brine Shrimp had the same overall dimensions as the boat shown here but did not fold. Like Weekender, it had asymmetrical hulls, and being under 14 feet wide it could be trailered with permits. In most states loads over 14 feet require very costly permits and much scheduling, leading and following vehicles, and so on. I figured the original Brine Shrimp could be towed to water in the spring and home again in the fall. Quite a few sets of plans were sold, but if any boat was completed I wasn't informed of it. Then came the brainstorm to design what was at first called Brine Shrimp II, which could fold to 8 1/2 feet and take the highway without permits. It seemed so much better that the original was soon taken off the market, and then the *II* wasn't needed.

It was Tom LaMers with his considerable trailering experience who produced the idea that makes this catamaran work. He pointed out that a hinge needed for folding doesn't necessarily have to carry any of the strain when a boat is in the water. This boat has the I-beams of all my multihull designs except Weekender. Heavy plywood (in this case 3/4 inch) forms the beam webs and goes down into

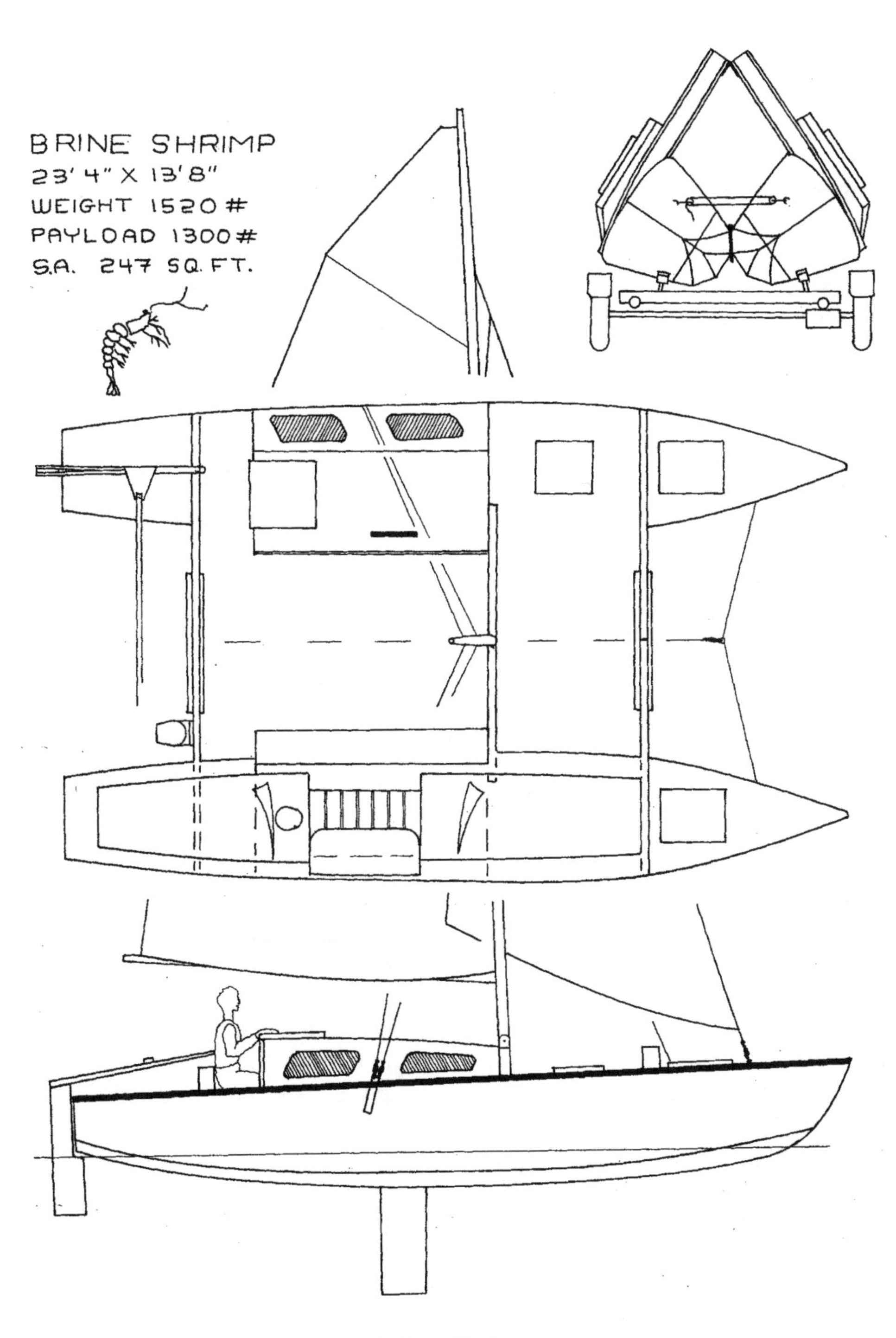


Brine Shrimp

the hulls, where it is glued and screwed to bulkheads. The beam caps are solid Douglas fir. Carol and I have sailed on boats with these I-beams since 1984 and have weathered countless gales at sea and one Force 10 storm, when we lay ahull with no sail up for 23 hours. The I-beams took it better than we did.

As can be seen from the drawing, the Brine Shrimp beams are hinged amidships, which is a catamaran folding method first used by Rod MacAlpine-Downie on the 20-foot-by-10-foot Shark daysailers that were in production a number of years. When Brine Shrimp is in the water the strain is taken by four 8-inch-by-4-foot butt blocks, also $^{3}/_{4}$ -inch ply with fir caps, bolted through the fixed beams so that the hinges are just along for the ride. The mast beam is a one-piece box section and is set and bolted in place after the unfolding.

The drawing shows the boat on the trailer with the butt blocks removed, but there is no reason why they shouldn't be permanently bolted to one beam, except that height of the load would increase from 8 feet, 9 inches to 10 feet, 5 inches. That's still low enough to clear electric lines, but maybe not all tree branches. Truck trailers are often a good deal higher than that, but they are better designed for whacking branches.

Retrieving a Brine Shrimp and preparing her for the road might take two people an hour or so, and the early steps might better be completed at a beach near the launching ramp rather than on the ramp itself. The tiller bar and rudders would be removed, along with everything on the inboard shelves. Next, the mast would be lowered—the mainsail has one full batten and I hope it won't be troublesome, but it allows the mast to be a couple of feet shorter and easier to lower. Two lashings that support the ply bridgedeck would then be untied from around the mast beam. That makes the deck springy although not unsafe, so it is best to have only one person on it from that point on. Mast and mast beam would be lashed on deck beside one cabin or other.

The butt blocks would be unbolted next, and the prudent skipper would have a few extra bolts, nuts, and washers on hand, as they are easy to drop into the water. Then the skipper would jump down into the water where the crew has been standing, holding the bows, and snap the vang tackle to U-bolts that are the chainplates of the inner forestay bridle; when he pulls on the vang the beams and bridgedeck rise and the hulls come together. At the last, before the hulls are quite touching, he would insert a piece of heavy carpet looped and sewn over a broom pole into the spot where the hulls would otherwise touch each other. The boat would be still perfectly stable, as waterline beam is about 7 feet, and it could then be towed around to the ramp and guided onto the carpet-covered bunks of the trailer.

Readying a Brine Shrimp for the trailer is a much bigger project than retrieving a Weekender, but you do get a lot more boat. She has almost three times the interior volume and payload and almost five times the deck space. She's about twice the weight and therefore takes about twice the time and money to build. Shelby Craghead, who built the Brine Shrimp shown in the photograph, worked even faster than Cliff Wade. I figure $4 and 40 minutes per pound of boat weight

Brine Shrimp hulls bolted together (Photo courtesy Shelby Craghead)

for these multihulls, with working sails and without engine, but of course those figures vary with the skills and finickiness of the builder.

This boat's trailerability influenced every phase of the designing. Obviously, the highly sloped cabin sides and off-center companionways are needed to fold the boat in this manner. Shelby made his cabintops 3 inches higher and his companionways even farther off center. He says the changes seem okay, and they certainly look okay in his pictures. The hulls could also have been beamier at the sheer if she didn't fold, with the greater flare that I usually put into multihull topsides, and then single-chine bottoms like those on *Dandy II* would have produced very little more wetted surface than the double chines shown here. However, the double chines also make it possible for a flat to touch a flat when folded, and over the road that's kinder to the structure than having a chine touch a chine. Freeboard could also have been greater if she didn't have to fold to 8 feet, 6 inches, and because the bow freeboard is only 30 inches, she needs a relatively long bow overhang to have sufficient reserve buoyancy for running down a sea in bad weather.

Despite all these compromises that trailering dictates, some people do need to trailer. Tom LaMers lives in Ohio, a good 75 miles from the Ohio River, which isn't itself the world's best sailing ground. He has a mass-produced telescoping cat that he trailers regularly to New England and Florida and sometimes even to the Texas Gulf Coast and to Lake Winnipeg. In California the natural topography and man's improvements of it leave very little sheltered water where boats can be moored or anchored, and a disproportionate number of builders who buy plans or study plans for Brine Shrimp have California addresses.

The East and Gulf Coasts and the Pacific Northwest have more sheltered water and more facilities for boats than are used. I was pleased when someone from Rhode Island asked for a nonfolding version. I drew up an extra sheet for the plans, showing a boat that would be quicker and cheaper to build, lighter weight, and with nicer cabins. That sheet now goes out with every set of plans, in the hope that some people who have been thinking of saving the cost of a mooring will come to their senses. The rigid version is still under 14 feet wide, so she could, of course, be trailered home occasionally, like the original Brine Shrimp.

On coming aboard *Dandy*, Tom LaMers always speaks wistfully of the space belowdecks and the long sight lines. His cat was designed when trailering beam could only be 8 feet, not 8 $^1/_2$ feet, so his hulls are less than 4 feet wide, and with other inches taken out such as tumblehome in the cabin sides, the sight lines athwartships are only about 3 feet. In fact, his hulls are so claustrophobic that he and Evelyn have concluded that they'd rather spend their nights in a tent on deck, and they use the hulls only for a chemical head and for stowage.

Brine Shrimp's cabin overhangs give her about 4 $^1/_2$ -foot sight lines athwartships, which is easier to take. Two people can sit comfortably facing each other across a drop-leaf table. The bunks are less wonderful, with 3-foot headroom only above one's head and one's body slithered down into a cavern with 2 feet or less of headroom. Again it is the trailering requirement that dictates the cabin's 7-foot length. The other hull is laid out the same, although the board trunk coming through it is enough clutter, and I wouldn't add a table. With a filler panel over the sole either hull could have a kind of double bunk where two affectionate people could share a pillow and perhaps some other pleasures.

Jim Wharram estimates that for every three sets of plans he sells, one boat is completed. I have heard from several people that they've started their Brine Shrimps, but only Shelby has sent pictures, and none has reported on sailing qualities. Wharram has been selling plans for forty years so I need to be patient, and meanwhile the sailing can be pretty confidently predicted from the numbers and from experience with other catamarans.

The waterline length-to-beam ratio is 9:1, and with a Bruce number of 1.12, hull speed should be exceeded readily. The Bruce number is the simplest of the weight-to-sail area formulas, in which the square root of working sail area (in feet) is divided by the cube root of weight (in pounds), which is boat weight with full payload. A value of 1.0 gives a boat of average performance. On the H-28, a much-beloved L. Francis Herreshoff design, the Bruce number is 0.91. This number is a good guide to performance in moderate to fresh breezes, when wave making is the biggest factor in resistance.

In light to gentle breezes skin friction largely determines resistance, and sail area is divided by wetted surface without any square or cube roots. *Skene's* feels the number should be 2.2, but in my experience narrow multihulls sail well in light air with a number of only 2.0. Brine Shrimp has just a bit more than that, with board down; the worst design problem with catamarans is keeping down the wetted surface of two immersed hulls, two rudders, and so on. It's the one way in which a trimaran is very much superior. The H-28 has about the same sail-to-wetted surface as Brine Shrimp, but I'll warrant that Shrimp would sail three miles to her two in light weather, and of course four to her two in a real sailing breeze. In really light going a cruising catamaran will not keep up with a good modern monohull such as a J/24 unless she has as big a rig as the monohull, which would make her a real handful on offshore voyages.

Yes, although Brine Shrimp folds for trailering, she is meant to be capable of crossing oceans, and the J/24 isn't. H-28s have done it, but the skipper needs patience. Other trailerable multihulls have made ocean passages: a MacGregor 38 catamaran completed a Bermuda Race at least once, and the Farrier folding trimarans seem to go wherever they like. We met one in the Azores, bound from America toward Germany. I have every confidence in Brine Shrimp's butt-block beam connections, and her stability numbers are better than those of *Vireo,* a Wharramesque catamaran in which we sailed transatlantic three times.

Payload would limit ocean crossing in this boat to two people, and even they would have to watch what they carried. However, boat weight was figured for khaya plywood, which can't be had now, at least not in America, and the logical second choice is okoume, which would weigh 150 pounds less. That could be added to the 1,300-pound payload.

The sailplan has two reefs in the main and one in the jib. The jib reefs 3 feet, 9 inches along the luff and 4 feet, 3 inches along the leech, which allows the jibsheets to be led to the same spot, reefed or unreefed, and saves having the jibsheet blocks on sliding tracks, which can be very annoying and time-consuming to shift when the wind is strong enough for jib reefing. A storm jib is also needed for ocean crossings, cut so that its sheets also lead to the same spot. We have carried storm trysails on multihulls but have never found much use for them except as self-steering sails.

There are two kinds of voyage you can make in any boat, and contrary to popular opinion, the difference is not whether you're within sight of land or not. Rather, the difference is whether you're within broadcast range of weather forecasts and close enough to shore to find shelter if bad weather is predicted. If you're not, you have to be prepared to take whatever nature dishes out. Carol and I sail direct from New Jersey to Block Island each summer, which is 181 nautical miles and puts us about 35 miles offshore, perhaps more if we're beating homeward. Through the usual smog we do not see the low coastline for many an hour; but the trip isn't an offshore passage because we can hear the forecasts and head in if need be.

One summer in Vineyard Haven the radio announced that an aged hurricane that had been milling around in New York State for several days was finding new strength and heading straight for us. We sailed *Dandy* over to Hyannis on Cape Cod and into a creek 150 yards wide and 700 long. Depths are not shown on the charts so no one dared follow us, but we had been there before. When the eye came past us with 60-knot winds, waves in the creek reached 6 inches in height with tiny whitecaps. Pull on the anchor rode was strong, and periodically I had to let out another few inches of rode, against chafe in the chocks. We ate and read and played cards, and in the evening the wind went down. Quite a sea saga, eh? We've had enough of such weather in open ocean. For us the fear of bad weather at sea is cumulative, and we're no longer willing to face the conditions that a Brine Shrimp could readily survive.

Most Brine Shrimps will be used as we use *Dandy:* never beyond earshot of weather radio. But both could take a storm if they had to, and in fair weather Shrimp will be a habitable boat with three places to hang out, and even on rainy days there are two refuges. I'd make the cabin interiors as different from each other as possible, with different color schemes and perhaps even with different furniture. But in good weather in port the bridgedeck would be the nicest place, with an awning and some folding chairs and perhaps even a folding table, though the cabintops could well substitute for that. There's plenty of space to stow bulky gear on multihulls, but it must be lightweight. We prefer a bridgedeck of cedar duckboards like *Dandy*'s, but that would have made folding more complex. Shrimp's plywood deck will be nice, too, and high enough off the water not to be slammed hard or often.

A *Dandy* dinghy could easily stow on the bridgedeck forward, and the plans allow an outboard of up to 5-horsepower, although less would be better. The transoms are too narrow to mount an outboard and still swing the rudders, so the motor must go on a sliding bracket on the aft beam, and it will cavitate in a chop because its prop will not be protected by the wave train of a hull. But where there's chop there usually is wind, and Brine Shrimp is primarily a sailboat. Turn off the motor and put up the sails. You'll be pleasantly surprised, and the competition will be chagrined. In most circumstances Shrimp will walk away from most sailboats under 35 feet whether they're sailing, motoring, or motorsailing. We're not in a race, of course, but look at how small they're getting behind us. What are they saying to each other? Isn't this fun?

# *Dandy* and *Dandy II*

Carol and I sailed twice to the Azores on the 27-foot Wharramesque cat and twice on the 28 $^{1}/_{2}$-foot Hummingbird trimaran. We returned once on each boat via Europe, Africa, and the Antilles, and on the second east-bound voyage sold each boat in Horta to local friends. Flying home after the second sale we had little trouble agreeing that we'd made our last ocean passage; now we wanted the habitability of a catamaran and didn't need more than a 1,000-pound payload for the inshore sailing that we had in mind. That would allow our new boat to be smaller and therefore easier to build and maintain and easier to handle, both under way and when launching at the riverside. Part of the time saved in building smaller could be spent on round bottoms, which are always more time-consuming than chines in a one-off boat.

Strip-planked mold for *Dandy's* bottom and one completed shoe

*Dandy's* hull bottoms are solid fiberglass, laid up over a male mold as shown in the photo. The parting agent was cellophane, which is often recommended, but unlike other plastics it does not stretch to conform to three-dimensional shapes. The slightest change in temperature or humidity puckers it, so it must be stapled down instants before the first layer of glass is put on, and even at that, one end of the mold may pucker while you're laying glass on the other end. Doing it again, I'd experiment with other kinds of sheet plastic (some of which dissolve in polyester resin). Because the mold was male, the pucker imprint on the glass at least is inside the hulls where it doesn't interfere with water flow.

The layup was one layer of 3-millimeter Coremat sandwiched between two layers of 1708 biaxial Fabmat with the outside filled and faired with q-cell putty. Coremat is Dacron, not fiberglass, with microballoons embedded in it. Unlike other cores it is saturated with resin, and it drapes over the form by the force of gravity, like glass cloth or Fabmat. Cured, it weighs just a little less than water, or 60 percent of what fiberglass does. I put on the first layer of Fabmat, followed immediately by the Coremat, which meant that when the first sanding came I was sanding soft Dacron, not prickly fiberglass. Fairly large parts can be laid up this way if you work by the clock, starting early in the morning with just enough hardener to kick the whole thing off as the temperature rises toward noon. However, after the final layer of Fabmat and the q-cells were put on, there was no avoiding the grinding of fiberglass.

In production shops, no effort is made to bring the glass neatly to the edges of the mold because it stretches and moves around while being wetted out. So it runs wild, and all layers are cut off at once after the resin has cured. With the *Dandy* shoes I made the mold a couple of inches wider than the part I wanted and a couple of inches longer at the transom. The desired line was marked on the strip planking, and just below it $^1/_4$ -inch holes were drilled about 18 inches on center. When the layup had hardened holes were drilled through the glass from the inside, a line was drawn with a batten, and each shoe was lifted off and cut with a carbide-bladed saber saw. The transom cut was drawn simply by laying the shoe over the wooden frames on the strongback.

The resulting fiberglass is little more than $^3/_{16}$ inch thick and would not be stiff enough for the nearly flat topsides of the hulls, but the bottoms have an 11-inch radius. Weight of the shoes per square foot is very little more than the $^3/_8$ -inch ply topsides. At the transition point between glass and ply, a wide stringer was notched into the frames and rebated to reconcile the differing thicknesses. The shoes were glued and screwed to the stringer and so was the plywood, and the join was covered with a layer of 8-ounce epoxy glass. It has given no trouble in eight seasons and perhaps 8,000 miles of use.

The frames go down into the bilges, and there is a 1-inch wooden keelson that was heaped with thickened epoxy just before the shoe was put on for glue-up. Another way to reinforce the bottom would be to put a piece of halved PVC pipe or even cardboard tubing inside the bottom after the hull was turned over and glass over it and out a couple of inches onto the hull bottom. But can it be done without dribbling resin around the interior? Such joins are known in the trade as "secondary bonds" and are weaker than the laminate itself unless done with epoxy resin.

Unfortunately I was hell-bent to put a biplane rig on *Dandy*—a mast on each hull. I'd been thinking about it for fifteen years, and Dick Newick's prediction ("good downwind in a Force 5") didn't deter me. After two seasons we thought better of it. Among other problems, the windward sail blanketed the leeward one through 60 degrees on either tack, and as most boats cannot sail within 45 degrees of the wind, that left only 150 degrees of the compass to sail in. Additionally—and contrary to what books tell us—the balance of a sailplan is more thwartships than longitudinal, and dropping one sail of a biplane rig makes the boat unmanageable. The chief reason that monohulls develop weather helm when they heel is that the center of effort moves out to leeward. One reason the mini-garvey doesn't is that when heeled the center of buoyancy and resistance is still nearly under the sailplan.

We also found that although the windward sail (there were no jibs) could be reefed readily, the leeward sail was hanging out over the water farther than it would in the narrowest monohull. The prospect of reefing that sail scared us so much that in our two seasons with the biplane rig we came home from New England down Long Island Sound, not direct from Block Island to Egg Inlet. The winter that *Multihull Voyaging* was being written, we were converting *Dandy* to a fractional sloop rig.

*Dandy* in Gloucester Harbor (Photo courtesy Susanne Altenburger)

These days even hot monohulls are likely to have fractional rigs, although it usually involves them in running backstays. Masthead rigs were popular for a very long time because they needed no runners and they allowed the largest sail area to be set on a given height of mast. The racing rules may also have favored masthead rigs, but I'll leave you to look that one up. The problem in multihulls is that jibs impose more compression strain on masts than do mainsails, and the light structure of a modern multihull flexes with that strain, which makes the jib luff sag. The photo of *Dandy* was taken in about 10 knots of wind, but you can see that already the jib luff is somewhat hollow. A jib can be cut flat so that it still draws with some hollow in the luff, but there are limits. When the luff gets too far to leeward the sail does more work as a brake than a driving force.

To avoid runners and still put some tension on the headstay, *Dandy*'s shrouds are swept aft 30 degrees in plan view, and some designers do it even more extremely. The mainsail and even the boom touch the shrouds long before the sheet is let out as much as it could be with an unstayed dinghy rig. A good vang keeps the sail off the shrouds for a while, and certainly every catamaran should have one unless it has a sprit boom like Weekender. But a vang to the mast butt is not very effective in a trimaran where the mast is stepped on the cabintop, and with such trimarans it is often necessary to vang the boom to the leeward float when off the wind. However, the sheet is not often eased out far on cats or tris because the asymmetrical spinnaker is set and the apparent wind is brought up to the beam or close to it. Thus the whole rig works together to give the best performance in these boats.

The masts on *Dandy* and Night Heron are 50 percent of the waterline aft, and even on Weekender and Brine Shrimp the forestays are well back from the bows. Lock Crowther explained to me that weather helm in multihulls is caused by having the rig too far forward where it presses down the narrow bows as the wind rises, shifting the center of resistance forward. On *Twiggy*, a famous 31-foot trimaran of his, the mast of the fractional rig was stepped 56 percent of the waterline aft of the bow, and he said that he'd like it farther aft, but the daggerboard would have to move aft with it and would interfere with the steering. As with a monohull dinghy, you cannot do better with the longitudinal helm balance of a multihull than to put the daggerboard directly under the center of effort of the sailplan and forget about cutting out the underwater shape, balancing it on a knife edge, allowing a certain factor according to what kind of rig it is, and all that dumb business. *Skene's* and Chapelle and a bunch of other books describe the procedure, but for my money it's an historical curiosity.

*Dandy*'s rig is not huge, and scantlings are heavier than some designers are now using. Despite that she is not a slow boat. Beating up into Vineyard Haven behind a racing fleet of Solings (a 27-foot Olympic keelboat) we had to ease the sheets to keep from passing them and perhaps spoiling their race. Near the head of the harbor where the wind was lighter, we sheeted in and headed up again, but they still pulled away, because our wetted surface was now hurting us.

Once we found ourselves in a Buzzards Bay regatta with the multihull class starting out five minutes ahead of a monohull class that included a J/27. That is one of the faster J Boats, and it had a crew of five, very high-tech sails, all kinds of winches, and so on. There were only two of us with three winches and sails of Dacron, not Mylar. It was about 3 miles to the windward mark, and in a decent wind it may have taken us forty-five or fifty minutes. Near the mark the J/27 overtook us and gave us the usual Corinthian salutations.

Rounding the mark, they set their spinnaker almost instantly for the broad reach to the jibing mark, while I was several minutes sorting out lines and getting our spinnaker to draw properly. Then we went past them like a rocket. I believe we were going twice their speed, and at the jibing mark they were indistinct on the horizon. When we jibed we struck our spinnaker for the shorter leg to the downwind mark. They jibed theirs with wonderful precision, and as far as we could tell their speed on that leg under main and spinnaker was about the same as ours under main and jib.

Of course, we would like to have that extra bit of speed upwind. Among other benefits it would save us hearing the comments of buffoons in J/27s. But the price comes pretty high in more ways than one. In a different race put on for multihulls alone, we set out from Vineyard Haven when small-craft warnings were flying. We were well reefed but still found the upwind legs trying. Back at the beach afterward I said to one of the competitors that I was pretty sure we'd had one hull out of the water several times. No, he said, I saw you at one point with both hulls completely out of the water and only rudders and daggerboard in. The other boats were stripped for racing, but we had aboard our whole household for the summer: food, clothing, books, pets. If the boat turned over it might have been righted unharmed, and our lives certainly were not at risk. But drowning pets for the sake of a trophy is not our agenda.

Another competitor in that fleet, racing the next year in Narragansett Bay, hung onto the spinnaker in his Farrier trimaran until the boat literally pitchpoled. The Coast Guard dragged it back to Newport inverted, and on the way the mast hit bottom and damaged the rig. The skipper was furious with the Coast Guard, but he might as well have been furious with Ian Farrier. In fact, it was his own fault, and his kind of attitude toward sailing is spoiling the sport, just as paid trips up Everest are spoiling mountain climbing. The purpose of extreme sports is to sell equipment, and there is no sport that suits extremism less well than sailing.

In *Multihull Voyaging,* to illustrate the differences between catamarans and trimarans, I drew a Hummingbird tri and a Hummingbird cat of the same 28 ½-foot length, which of course had a nice big bridgedeck and much more accommodation. That was the boat, never drawn in any more detail than a study plan, that created the most interest; and I had a deluge of letters about her, asking whether complete plans had been drawn yet. When *Dandy*'s rig was sorted out, and we had a few seasons to appreciate how very pleasant a boat she is, plans were drawn for *Dandy II,* enough bigger to have an ocean-going payload and more space in every dimension.

The hulls are single-chined like the Hummingbird tri's because not many builders want to mold fiberglass shoes. An overall length of 28 ½ feet didn't seem necessary, and at 26 feet, 9 inches she's what I consider a big boat. If you build her

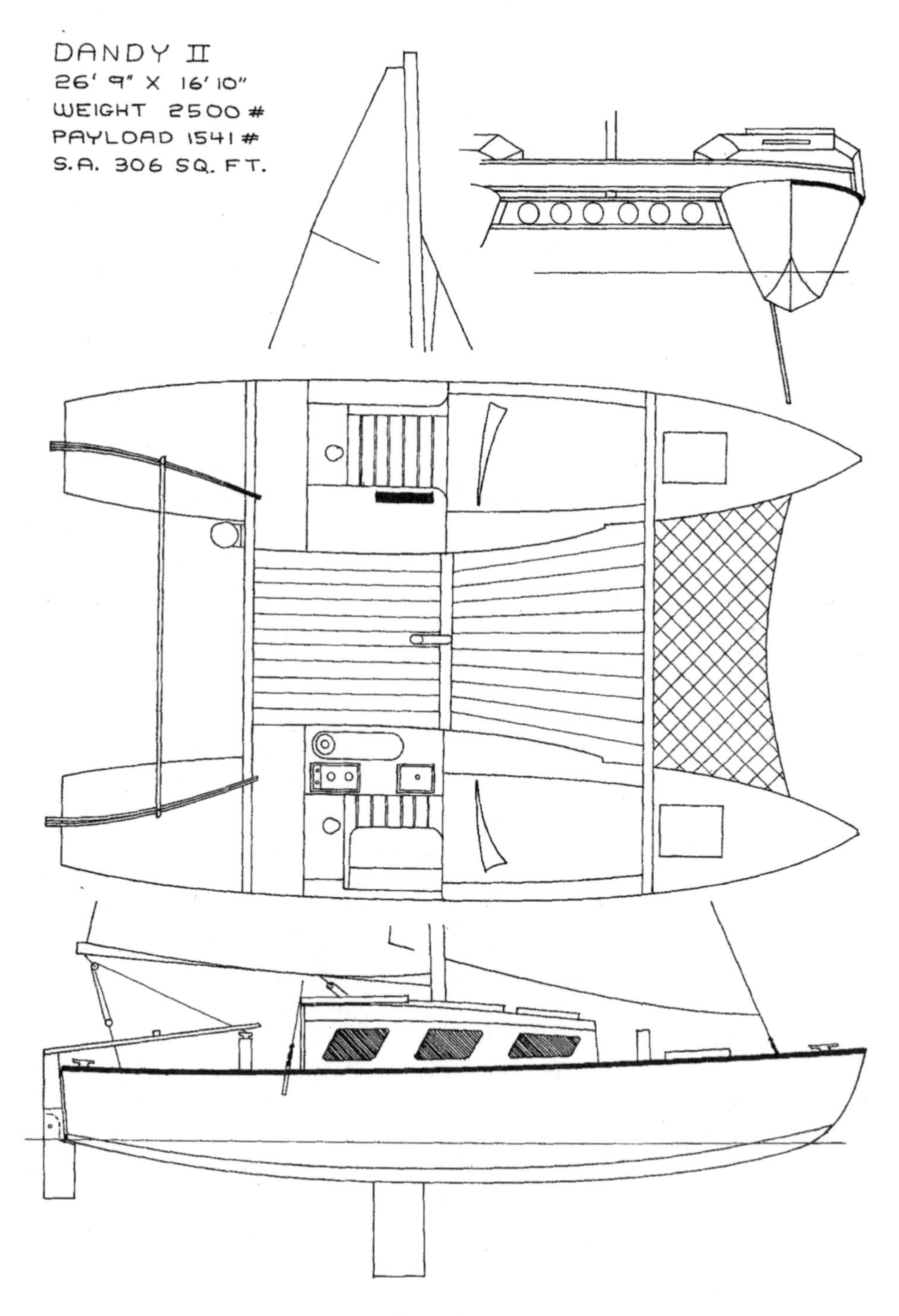


Dandy II

yourself, you'll reach the same conclusion. She is indeed an habitable vessel and grand enough to support a few superfluous beautifications such as the hollow sheer line, which looks more boaty to all of us than a straight sheer. I'm not convinced it's one bit better functionally, unless the owner is planning to haul fishing net.

The *Dandy II* design is new, and no boats have been completed that I know of, but her performance should be very like *Dandy*'s. Both have hulls 10:1 on the waterline, sail area to wetted surface of just over 2.0, and a Bruce number of 1.13. The chines of *Dandy II* will produce some turbulence and resistance, but not to the degree that monohull chines do. Tank testing with two otherwise identical models would be required to pin down how much that resistance amounts to. But *Dandy II* is 2 feet longer on the water, so at any given speed (for example, 6 knots) she will have a lower speed-to-length ratio, and therefore less resistance. My guess is that if *Dandy* and a *Dandy II* left port one morning, bound for the same destination, in the evening they'd still be match racing. Naturally I hope to find out one day.

On our Wharramesque catamaran the bridgedeck was only 4 $^1/_2$ feet wide because the whole boat was narrow, compared to my designs or Jim's more recent ones. Compared to what you get on a 40-foot monohull, 4 $^1/_2$ feet by 14 feet is a palatial lounging space, but it seemed a bit like a corridor and when one crew sat in a chair athwartships, he had to draw in his legs when another crew wanted to get by him. Brine Shrimp has a 5-foot bridgedeck, chosen because the space inside the cabin overhangs seemed more valuable than a wider space on deck. Both *Dandy* and *Dandy II* have 5 $^1/_2$ -foot bridgedecks, and that really is the ticket. Not only is there lounging room, but there's circulation, as house architects like to say. Perhaps in your house the livingroom is between bedrooms and kitchen and it seems there is no corridor. Actually there is one: a space in the livingroom that you can't furnish because it's needed for circulation.

The masts of the *Dandys* are supported by a third I-beam that passes underneath the bridgedeck to leave that space clear for work and lounging. The beam bottom is only 18 inches above water, and in a 27-foot catamaran a bridgedeck that low would slam badly. Such slamming can destroy certain mass-produced cats that have cabins on the bridgedeck and are struggling for headroom inside them. But *Dandy*'s mast beam is relieved with big round holes, so there's little surface for the waves to grab. Like the steel trusses that you see in the construction of big buildings, holes in the web do not weaken a beam.

In nearly all Wharram's designs bridgedecks are duckboards instead of solid ply to lessen the impact of occasional slams, and it does help. *Dandy*'s bridgedeck is copied from him and is white cedar $^3/_4$ inch by 5 inches and $^1/_2$ inch apart. Being unpainted, the boards give some traction even when wet. The cedar slowly oxidizes in sunlight, and if it isn't scrubbed with a stiff brush every few months, your feet will start making gray footprints when you go from cedar to the painted hull decks. The cedar weighs no more than $^3/_8$ -inch plywood when dry but does soak up more water, even from dew once it has been wetted with salt water.

The layout of both *Dandy II* hulls is shown on the drawing because they are very different. The port hull has a bunk 46 inches wide. It is 24 inches above the sole and must be hopped up on, not sat down on. The starboard bunk is meant for a dinette seat as well and so is lower and narrower. The aft decks are over 8 feet long and could possibly have bunks under them, but headroom would be only 2 feet. And on a rainy day in port, what are five or six people going to do on this boat except pick fights? The big chart table in the port hull is pierced by the daggerboard trunk, so it's less useful than it looks but is a fine place for stowing things that should be taken ashore and thrown out. One plans buyer expects to put stowage cubbies behind the trunk.

The starboard hull seats four comfortably, and there is a two-burner propane stove with a self-draining well beneath it. The propane bottle is in another self-draining well in the cabin overhang, and the copper connecting tube goes out one drain hole and in the other, so it's hard to imagine a propane debacle. This outside well is also the best place to store gasoline bottles and other combustibles. The sink in the drawing has a drain but probably shouldn't. A place to put dirty dishes is nice, but they should be washed in sea water on deck. A 30-gallon tank of fresh water won't wash many dishes, but will consume 20 percent of the payload of *Dandy II.*

Fooling around with the accommodation plan is what people like to do when building a multihull, and within limits there's no harm in it. "There's so much interior volume, and surely I can find a better use for it than that designer bozo." Maybe so, but stringers should be kept about 12 inches apart, and weight of furniture matters, as well as what goes on it and in it. One guy who bought plans wanted to stretch the whole boat, which would have lengthened the bunks from 6 feet, 8 inches to 7 feet, 6 inches and put the frames farther apart in the critical spot 33 percent of the waterline aft, which takes the most pounding in heavy weather.

For auxiliary power 5 horsepower is the maximum that this catamaran can use. *Dandy* is 27 percent lighter, and 2 horsepower drives her fine. A friend of ours recently bought a ferrocement Tahiti ketch. The replacement value of her single-cylinder diesel engine was over half the price he paid for the whole package. Bringing her home through the Chesapeake and Delaware Canal he discovered that the diesel wouldn't push the boat against the current, and therefore he ripped it out and threw it away. He's thinking of four cylinders, but they won't push that hull appreciably faster. These are sailboats, although the Tahiti ketch is a bad sailboat, and engine power giving square root of the waterline, which would be 5 knots in *Dandy II,* is all we can expect without degrading performance under sail. Occasionally one may just have to wait out a canal tide.

*Dandy II* is barely big enough to have an electrical system, for those who crave it. Shelby Craghead means to install one in his Brine Shrimp, and I hope the wires will stand the folding. Some 5-horsepower outboards do have small alternators. The smallest useful wet-cell battery weighs about 50 pounds, and battery box, wiring, and breaker panel may weigh another 10 pounds. Electric lights are what people want most, and probably they weigh less than kerosene lanterns, not to mention the weight of kerosene. But the alternator of a small outboard typically produces 5 amperes, while a GPS consumes 0.25 amps and as much as 2 amps if it shows us charts in color. A VHF radio uses 0.5 amps for listening and 5 or 6 amps for talking. So if a battery stored and gave back half the electricity put into it (and what battery is that efficient?) the motor would have to be run a third of the time just to power the GPS and the VHF on standby. Once again, it's worth remembering that we're talking about sailboats.

The rudders shown in the drawing kick up. The stocks have metal plates on each side and a wooden blade within. The blade is pushed down with a stick, and a tapered dowel is driven through a hole in metal box and blade. Hitting bottom or flotsam breaks the dowel, and this is a relatively simple system to build, but in a seaway getting the blade down again and the remains of the old dowel out and new dowel in can be quite a chore. The boat will steer with the rudders kicked back, but the helm is very heavy, because the rudders are now long and low-aspect.

A better but more complicated system can be made with a longer metal box and a rudder pivot bolt higher up, near the tiller. A rope goes around the box and the back of the blade and up to a cam cleat mounted on a piece of fiberglass about $^{1}/_{16}$ inch thick. When the rudder hits something, the fiberglass bends, and the cam cleat releases the line. Once past the flotsam and into deep water, the rope is pulled, and the rudder returns to vertical. The trouble with a catamaran is that you have to make two of these damned things.

Despite the duplication of many parts of a catamaran (and the duplication of some accommodation features, too), Carol and I wouldn't think of any other kind of boat for cruising. Although not as fast as a trimaran, a cat is fast enough for us. Each multihull that we've had has suited us better than the last, and we've kept it longer. That should be happening to you too, no matter what kind of boat you prefer: daysailer, kayak, runabout. Learning about boats is fun, and it also increases your fun out there on the water.

# Night Heron

Forty years ago, when Americans and Europeans first became interested in multihulls, it wasn't understood what made a good tri or a good cat. Arthur Piver more or less invented the trimaran, and others who quickly followed him didn't stray far from his proportions. Main hulls had a length-to-beam ratio of as little as 5.5:1, highly flared topsides, and bunks on the wingdecks even in boats under 30 feet, so they had tremendous interior volume but moved well only with strong wind aft. Catamaran designs were adapted from Pacific Ocean working craft. Hulls tended to be very narrow and closely spaced, practices forced on the Pacific islanders by the materials available to them. So the early trimarans were habitable and stable, while the early catamarans were fast and tippy. Nowadays we understand that a good cat is more habitable, while a good tri is faster.

Night Heron is only 10 inches shorter than Brine Shrimp, so they can be compared pretty exactly:

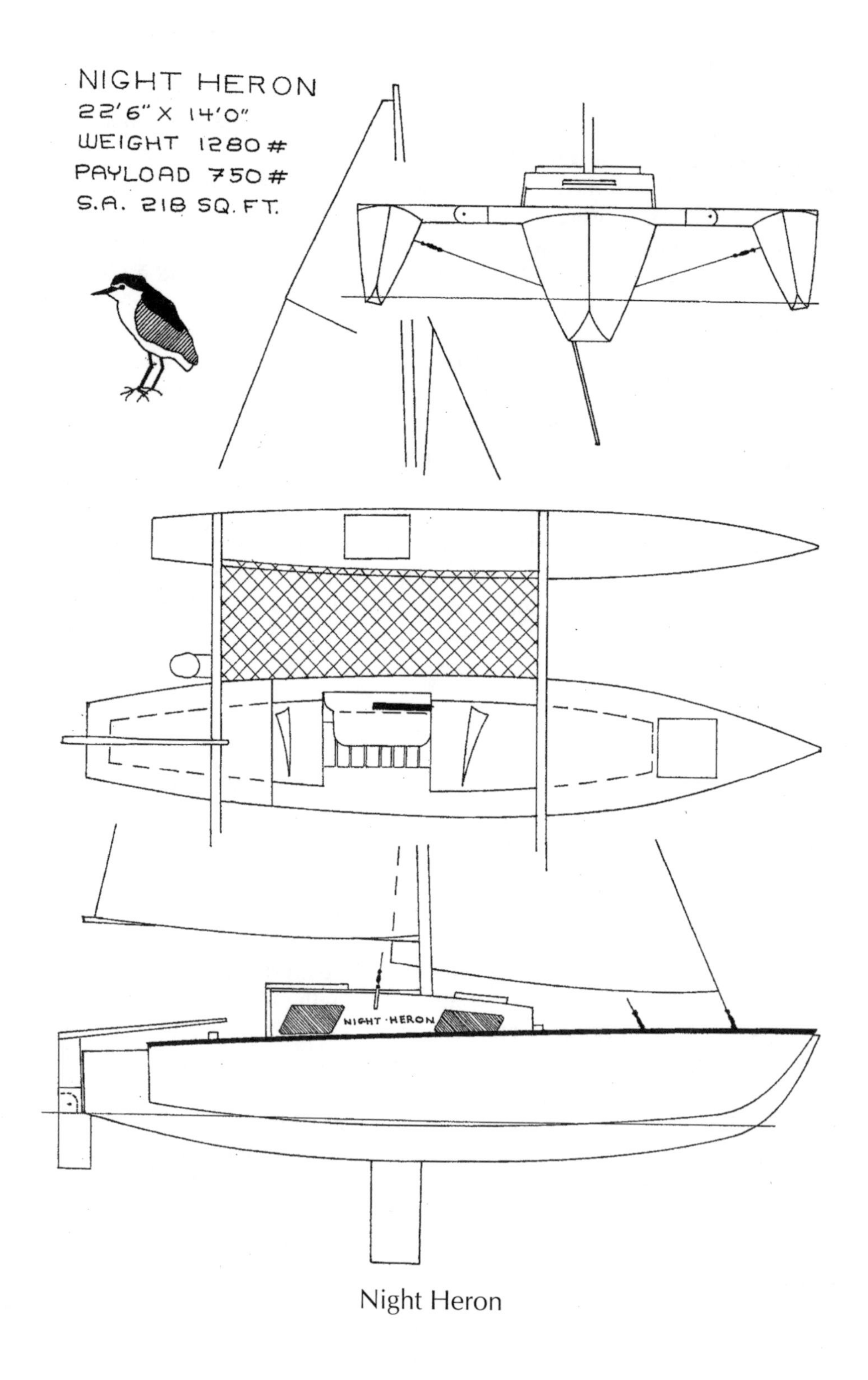
NIGHT HERON
22'6" X 14'0"
WEIGHT 1280#
PAYLOAD 750#
S.A. 218 SQ. FT.
NIGHT HERON

Night Heron

| | Heron | Shrimp |
|---|---|---|
| Payload | 750 lbs | 1,300 lbs |
| Interior volume | 183 cu ft | 337 cu ft |
| Bruce number | 1.17 | 1.12 |
| Sail area-to-wetted surface ratio | 2.47 | 2.04 |

The first two numbers indicate habitability. In both boats the interior volume is what's available when you pop down the companionway and does not include forward holds or trimaran floats. It turns out that Heron could have had 20 cubic feet more space inside if I had dared to draw cabin overhangs. The recently completed prototype weighs 100 pounds less than was expected, which would give thwartships sight lines belowdecks of nearly 6 feet. That space could have been used only for shelving, not for more bunks. Anyway, 850 pounds is little enough payload, and it still precludes ocean crossings, except singlehanded.

Brine Shrimp's habitability stands out more than these numbers indicate: she will sail to windward at no more than a couple of degrees of heel, while in a stiff breeze Heron may heel 15 degrees. Her bridgedeck will be more pleasant in port than Heron's wingdecks, because when crew congregates on one side of a trimaran the float sinks and the boat again heels. With two people in the main hull—sleeping or awake—a trimaran also bobbles back and forth between her floats in wakes or other waves, with consequent splashing. That happens under way as well as at rest. Dick Newick once told me he thought that trimaran motion was easiest of all, and in *The Cruise of the* Alerte, E. F. Knight similarly argued that a monohull with all her ballast inside rolled deeper but easier than a boat with a ballasted keel. Let us not deprive those guys of their pleasures, but for easy and minimal motion give me a wide-stance catamaran. It's next best thing to being on land.

The last two numbers foretell performance, but I don't think the slightly higher Bruce number of Heron makes her faster in strong winds. The main hull is only 8:1, and although she lifts it out far enough to be 9:1 or better on a reach, she doesn't on a run. In addition, the simple dory-shaped hull creates some chine turbulence, although not as much as a fatter monohull dory. The flat bottom makes for easy building and trailering and gives the displacement below waterline that a chined shape couldn't have unless it were wider or deeper or both.

In 1991 Carol and I sold our 28 $^1/_2$-foot trimaran to José Fraga in Horta, and he has put on her the speed-measuring equipment that we never had. She also has an 8:1 main hull, and the best speed he's reported is 14-point-something knots. I would think that all the boats in this chapter except narrow Weekender could come close to 15 knots on occasion, although none could average it for an hour, let alone for a day. Speed numbers are so freely thrown around and so wildly exaggerated these days that it's hard to remember that 15 knots, or 17 $^1/_4$ miles per hour, through the water is very fast and not much fun, except on special and brief occasions. On the narrow, winding Tuckahoe River where we have tried out the first Night Heron, we haven't had the conditions to get her up to top speed, let alone to measure it.

The sail area-to-wetted surface ratio is the telling comparison. Heron's is about 20 percent greater than Shrimp's; as a result, in light air she'll go a good 50 percent faster. You must sail on a spritely trimaran before you believe how well these boats move in the kind of breeze that's most pleasant to sail in. You can hardly feel the wind on your cheek, but looking at the wake you see that you're doing 3 or 4 knots. By moving through the air, the boat makes wind and then sails on that. Long ago, epoxy salesman Meade Gougeon entered a daysailing trimaran in a race that featured many different designs and completed the course in winds so light that the other boats could barely cross the starting line. A commentator remarked that cigarette smoke was going straight up.

Because she has accommodation, Night Heron is not as quick in light air as Gougeon's tri, but she is quick enough to amaze and delight. Carol and I are accustomed to sailing *Dandy*, and we like her very well, but an afternoon's spin on Heron sets us wondering whether we've yet found the perfect boat. In addition to speed, Heron is as maneuverable as a monohull racing dinghy. Catamarans are rightly said to tack slowly, although with a good daggerboard and rudders they *will* tack without backing the jib, even in a seaway, until reefs are necessary. The problem is that one hull must swing around the other, describing a bigger arc, and the ends of both are to some extent going sideways through the water. Ackerman linkage on the tillers helps with that problem because it causes the outboard rudder to turn at a lesser angle than the inboard rudder, just like the Ackerman linkage in automobile steering. But a cat still doesn't tack quick.

Once we were beating down the south coast of Cape Cod on our Hummingbird trimaran. It was nearly time to reef, and when we tacked I saw the new windward upper shroud fall inboard against the cabin. A clevis pin had dropped out of the turnbuckle because we were using cotter rings in the clevis pins, not cotter pins—rings are handy, but they should be taped in place to prevent unwinding. I knew that if the sails filled on the new tack we would lose the mast, and I shouted to Carol at the helm to tack back. The tri had enough momentum to do so, and I was able to find a new clevis pin and ring before we ran out of sea room. On a catamaran the story would have had a less happy ending.

A trimaran is easier to trailer than a cat because you are folding or telescoping light floats rather than heavy main hulls. Heron's floats weigh less than 200 pounds because the whole boat is okoume ply with 9-millimeter decks and bottoms and 6-millimeter topsides. The floats fold in neatly against the main hull. When unfolded, the $^{1}/_{4}$ -inch stainless steel hinge plates are stressed only in compression because the upper shrouds take the downward strain of the floats (which won't exceed their weight unless they are crammed with stores) and the wires under the wingdecks take the upward strain, which just before capsize could be nearly the total displacement of the boat. The many wires and turnbuckles are one of the costs of trailering Night Heron.

We have launched the prototype only with rollers from our muddy beach, but for trailer launching the drill is to loosen the hinge bolts and swing the floats far enough apart so that when they touch the water they open farther, not close again. The wires under the wingdeck should be pre-set, and thus will not need adjustment. Like Brine Shrimp, Heron will need nearly 30 inches of water before she floats off, and as most launching ramps have a slope of 1:10, the tires of the towing vehicle will be in the water by the time the boat reaches that depth, although I hope the axle won't. Then the rig is erected and the hinge bolts tightened. Launching Heron will be a good deal less trouble than launching a cat. When retrieving the boat, not everything will need to be cleaned out of the main hull because it is always upright, although more weight in her will require more water for launching and hauling and will put greater strains on her bouncing over the road.

Most trailering trimarans fold this way, tucking the floats under the main hull flare. Heron is about as big a tri as can use this simple system with a single hinge pivot, but with a more complex multi-pivot system Ian Farrier is designing and manufacturing trailerable trimarans as long as 31 feet. You don't get proportionately more main hull breadth as the boat gets longer because the whole thing must still fold to highway width, but you do get lengthy accommodation with an aft cabin on some models.

One evening we were standing at a launching ramp in New England with Jan, the other Gougeon brother. A Farrier trimaran was launching and unfolding before our eyes—and hurriedly, because the skipper was thinking about a timed drawbridge just downstream. At every step—and there weren't many—Jan was wreathed in smiles, saying, "Isn't that neat! Isn't that neat!" as if he had never seen it done before. Indeed, it was neat. Farrier's tris also sail very well, and they have Spartan but sufficient accommodation for those who find molded plastic homey. They are not expensive for what you get, but they still cost quite a bit of money and are quite a big commitment and responsibility, no matter how deep your pocket.

Night Heron is built just like the three catamarans discussed in this chapter, on a strongback with stringers over frames. Plywood boatbuilding is still evolving because the material isn't any older than fiberglass. Truly waterproof ply became possible with the invention of truly waterproof glue during the Second World War. At first, plywood was simply substituted for planks over the same closely spaced frames that planks need, to keep them from working against each other and leaking. Then designers figured out that with plywood, frames could be farther apart, perhaps as much as 2 feet. The Wittholz catboats date from that era, along with a good many other designs still being sold today. Typically the frames are oak, as they would be for board planking, because oak holds fastenings well, and glue was not trusted to replace or even supplement screws. In fact, certain resins in oak keep it from holding glue.

Stringers-over-frames was the next step, and by now all framing has become fir or mahogany, which do hold glue. It came to be understood that frames could be 3 to 4 feet apart if longitudinals about 12 inches on center were used. It was (and is) recommended that planking be screwed to longitudinals only because they are said to flex between frames, which could cause a screw in a frame to leak or to pop its putty covering. In practice I've never seen that happen, but maybe it can in high-powered runabouts, and at any rate screwing the ply to the longitudinals seems to be sufficient.

The taped-seam technique discussed in chapters 1 and 2 is the latest phase in plywood boatbuilding. It does have advantages and is the best method for many hull shapes, but in the long, narrow hulls of a multihull it has limitations. If the ply is tortured throughout its length, it is stiff enough, and the Gougeons have done that successfully a number of times. Typically the length-to-beam is 20:1 (at the sheer, not at the waterline), which doesn't leave much room for accommodation. The Gougeons frankly say, "The real design problem lies in achieving enough displacement for a given length of hull."

The floats of Heron are 20:1; they could be tortured taped seam, at some saving of weight. But the bulkheads to which the beams are attached would have to be aligned with machine-shop precision, and because the main hull cannot be taped seam, the builder (perhaps building his first boat) would have to work with two different methods. As is, the three hulls are made with the same strongback and cross-spalls for the frames, which are never moved, so beam alignment is assured. Only forward and aft of the beam-attaching frames are the cross-spalls moved after the floats are finished and before the main hull is begun. That is the recommended building sequence because novice builders learn the technique on hulls that are less complicated and experienced builders who find the floats a bit tedious to make save the most interesting part until last.

The daggerboard trunk is made up as a box with the inside surface fiberglassed before assembly, because it is never possible to paint more than a few inches down and up into it. Our Hummingbird tri had trunks in the floats and one board that was shifted when tacking, because it would work no better than a leeboard when in the windward hull. That boat was most often used for long passages, and shifting the board every day or two did not seem onerous. José Fraga now uses the boat mostly for afternoon races round the buoys, and he soon made a second board. As he does not carry cruising gear, the weight of the second board is acceptable. Boards in the floats are nice because they don't intrude into the accommodation.

It is expected that most Night Herons will be used inshore and tacked often, so a single trunk in the main hull seems better. As with the catamarans, if Heron's board were full length and raised downwind, the boom would hit it when jibing. Instead, it has a 3-foot handle of copper tubing that pivots to lie flat on the cabintop when the board is raised. There are two holes in the board through which a dowel can be slipped: one to hold it flush with the bottom sailing downwind and the other, 15 inches farther down the board, to hold it out of the water entirely when at anchor.

Usually when fitting a deck or cabintop, at least $^{1}/_{16}$ inch of ply is left overhanging to be planed off after the glue has set. Often during the planing off, though, the corner of the plane blade catches on the topside or cabin side and gouges it. That can be avoided by attaching a small L-shaped shim of thin aluminum plate to the side of the plane, using a small clamp. The other leg of the shim rests on the plane sole, just aft of the blade corner that might do the gouging. I have a larger and somewhat more elaborate shim and clamp that attaches to an electric plane when I am doing heavier work.

Night Heron's main hull has a copper vent under the tiller like the one described in chapter 6. The collision bulkhead forward has enough holes in its top edge to give about 7 square inches of ventilation. The aft side of the hatch to the forward hold is propped up on anchor, so that the hull has through ventilation.

Night Heron (Photo courtesy John Guidera)

Contrary to what we would think, when a boat faces the wind the air wants to circulate through it from aft forward. It takes a good deal of force—such as a large windscoop—to reverse that natural flow, and of course that's difficult to do without bringing in rain as well as air.

Venting trimaran floats is more difficult because in a decent breeze some spray flies past the leeward one, and in a real seaway they may even be submerged momentarily, although each one has enough buoyancy to support 150 percent of the boat's total displacement. Bound for the Azores on our 28 $^1/_2$ -footer, we noticed after several days of weather that wasn't horrendous that the boat was listing; we found one float nearly half full of water because we had failed to dog down a float hatch. Thank God the weather wasn't worse! For a good hatch seal, by the way, nothing beats the tubular vinyl gasketing sold in hardware stores for dealing with drafts around leaky old windows. The tube is $^1/_4$ -inch diameter and has a $^1/_4$ -inch flange that you tack down to the hatch coaming. The tube part goes outward, and you miter it at the corners. *Dandy*'s gasketing shows no sign of wear and has given no trouble in nine years.

On the prototype Night Heron the floats are ventilated by propping open the hatches when at rest. They are big hatches and provide about 15 square inches of ventilation, which isn't as good as 7 square inches of through ventilation, but it's some help. Because it is presumed that the boat is usually on a trailer with the floats folded and the hatches vertical, the hinges are on the inboard sides of the hulls, which makes them hard to use when the boat is under way. The spinnaker and anything else likely to be needed when sailing should be kept in the forward hold of the main hull or on a bunk.

Despite these trailering complications, this little boat is a whole lot of fun on the water, and there's enough space in the cabin for two people to be very comfortable. As the bunk cushions are 34 inches wide at their ends, even four people might sit out a rain shower belowdecks, although the Coast Guard insists on 18 inches of seat width for each passenger when licensing headboats. With the 2-horsepower outboard borrowed from *Dandy*, Heron goes considerably faster thanks to her lighter weight and smaller wetted surface and may reach her hull speed of 6.1 knots. *Dandy* must settle for 4.5 knots with that engine.

Because I had an order for another boat and a contract for this book, the prototype Night Heron was sixteen months abuilding, but when photography day came we had plenty of help. Bob and Jean Grimson had come upriver for a visit on their way from the Bahamas to New England aboard *Meander*, their 37-foot ferrocement ketch. She's a lovely boat and a nearly perfect liveaboard, although I wouldn't have encumbered her with gaff rigs on both masts. Nevertheless, she moved well the day I sailed on her.

Because the Tuckahoe is small water for a boat the size of Heron, I needed to be constantly in the companionway to handle jibsheets while Carol sat on the aft beam and steered. That put the boat badly out of trim, so Jean Grimson sat on the foredeck. Bob's weight there would have been even more useful, but he and two other photographers were running *Puxe*. Jean had never sailed before she married Bob ten years ago, so she had no experience with good daysailers or other lively unballasted sailboats.

It took nearly an hour for each photographer to run a roll of film through his camera, and before it was over Jean was literally screaming with delight. She just had no idea that sails could make a boat accelerate the way a throttle does a motorcycle. Bob was amazed that Heron didn't have to gather way on each new tack. She came through the wind so quickly that—if jibsheets were smartly handled—very little momentum was lost. During the postmortem around the kitchen table, I rather unkindly pointed out that part of the reason was that we were tacking in 90 degrees. "It looked like less," Bob said. "That's what you call *going to windward.*"

The Grimsons have been living aboard since they were married and have sailed *Meander* at least 40,000 miles. Most of their sails are still the originals, although even they can see that the end is fast approaching. Our own experience is that if a cruising boat (Bermudan rigged) with new Dacron sails tacks in 90 degrees, and if the sails are scrupulously kept out of the sun when not in use, after 5,000 miles the boat tacks in 100 degrees, and after 15,000 miles the sails are rags. Friends who use Mylar sails tell us that the sails hold their shape all their lives, but when they go they simply explode: there's no repairing them.

A trimaran is a different kind of boat, and cats and tris are too often lumped together in people's minds. Every summer that we sail to New England we hear some landsman on a dock seriously instruct his child that *Dandy* is a trimaran, and even experienced monohull sailors say, "Oh well, what's the difference?" The difference is very great. In a gust of wind, a cat accelerates like a tri, but it does not come through the wind with its momentum intact, and it cannot ghost like a tri.

One early September afternoon we were becalmed off Beach Haven Inlet on our Hummingbird tri, bobbing around among fishing boats. Ruth Wharram was with us. We had 15 miles to go to Great Egg Inlet and 4 hours to make the tide. Carol was due to start teaching the next day. The wind came in light from ahead.

I got us moving and gave the helm to Ruth, but she was no good at it. She can steer a good compass course, and she is certainly the best seaman I've ever known, more at home on the ocean than on land. But her experience is with big, heavy-laden catamarans, which need nearly as much wind as *Meander* to move well.

Carol and I took turns nursing the tri away from Beach Haven while the fishermen watched us and hung onto their beers in the slop from the last wind. By the time we could see Atlantic City the wind had freshened until any decent sailboat could go to windward to some extent, although not the way we were going. We did make the tide and carried it up to Tuckahoe, and Carol did get to school in the morning.

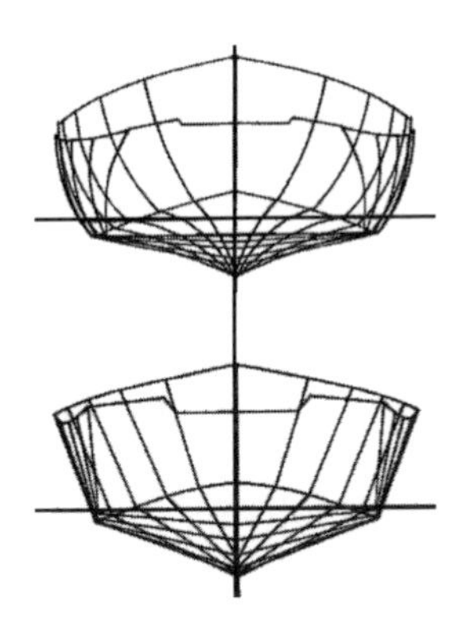

# 6

Details

# Details

*Details are the giant's fingers.*

—John Updike

## Spars

In building wooden spars, weight matters even more than it does in hulls; spars are high in the air and have a great effect on the center of gravity and the motion of a sailboat. In choppy conditions running spars—booms, gaffs, yards—shake the wind out of sails directly in proportion to their weight. And booms are virtually certain to hit one noggin or another sooner or later, again with impact proportional to weight. So the weight of spars needs to be pared to the last ounce. As discussed in the Melonseed chapter, I find Douglas fir the best wood for making spars; each stick should be carefully selected for weight and straight grain.

For short unstayed rigs such as gaff or sprit, solid masts are best; mast diameter should be 0.015 times length. All spars should be laminated against warping from at least two pieces of wood, or you can rip a plank and turn one piece end-for-end to achieve the same result. The warp-prone "grown sticks" of ye olde time—with their knots, checks, and other entertainments—should not be considered. Small solid masts should be tapered to half the diameter at the head that they are at the partners while they are still square. Lugsail and spritsail tapers can start halfway up, and a gaff taper should start above the gaff jaws.

To round a spar, the corners are cut off at a 45-degree angle, perhaps with a table saw or Skilsaw, to a line drawn 29 percent of the diameter from each corner. The mast is then eight-sided; to make it sixteen-sided I like to sit it on a couple of blocks that have V cuts of 157 degrees so that a plane held horizontal can be run over the corners. Very large sixteen-sided spars can then be rounded with a sanding belt turned inside out and propelled by a drum that is chucked into an electric drill. Smaller spars are best rounded by first giving them roughly thirty-two sides with a block plane and then finishing with a jitterbug sander. If the mast is to be square at the base, the transition to round can be made neatly with a spoke shave.

Short-rig masts that are 15 feet or longer should be hollow. Diameter has to increase 10 percent, say from 3 inches to 3 $^1/_3$ inches, but wall thickness can be 15 percent of diameter, or $^1/_2$ inch in this example. A solid tapered mast 16 feet, 8 inches by 3 inches would weigh 20 $^1/_2$ pounds, and theoretically a hollow one 12 $^3/_4$ pounds. Not all of that 7 $^3/_4$ -pound difference can actually be saved because the hollow mast must be solid at the head and beefed up where hardware, partners, and step are, and it probably isn't perfectly round inside. But more than 5 pounds can be saved, most of it high off the deck; and it is definitely worth the trouble.

Hollow round masts often are octagonal inside; the traditional way to make them is with the eight-stave system shown on the left in the hollow mast sections illustration. However, aligning all these bits is pretty tricky. Usually they are glued up in two half-circles before the final gluing. When designer James Wharram was building his present 63-foot catamaran he showed us the masts being made (center drawing), and I have since seen this interlocking method described in magazines.

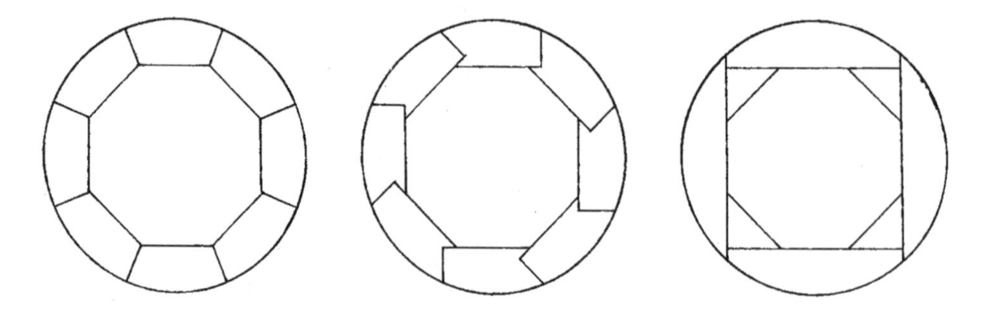

Sections of round hollow masts

It requires extremely precise table-saw work, which the equipment I have just isn't up to; but the glue-up certainly would be easier, with hose clamps used to hold it together.

The round hollow section that I use is shown in the right-hand drawing. While the four main staves are still rectangular, the triangular pieces are ripped and glued to the narrower staves, with brads to hold them while the epoxy goes off. On a catboat mast with a $4\,^1/_2$ -inch diameter, I recently hollowed the main staves to make wall thickness more uniform and save a bit more weight. I used an old wooden-bodied smoothing plane that I bought cheaply at a flea market because the nineteenth century was just as tool-crazy as our own. The plane body was rounded in section with an electric plane, and the blade was rounded to match on the grinder. Such a plane cuts only one radius, so you need quite a few if you make a lot of different-sized masts.

These days, bigger sailboats with standing rigging are usually Bermudan-rigged—and a good thing, too. Big sprits and lug yards are very troublesome to handle, especially in rough weather, because as soon as you start lowering them, neither end is attached to anything. On gaff rigs big enough to have shrouds, the sail twists a great deal as soon as the sheet is eased, and unless there are spreaders to prevent it, the gaff jaws are soon rubbing on the shrouds, which can chew right through them in a very short time. That is why the sheets of old three-and four-masted coasting schooners were eased very little, even when sailing downwind.

Dave Gerr's *The Nature of Boats* is now replacing *Skene's* (which is at long last being reprinted) as a design reference, and it isn't as thorough, but some of its formulas are easier to use. Presumably Gerr's mast specifications are for aluminum; he says that width should be 1/90 times length, and I have found that foolproof for hollow wooden masts too, even when the boat is as stiff as a multihull. With the six-wire rig of my multihull designs in chapter 5, the mast is nearly as well supported fore-and-aft as athwartships, and section length need only be 20 percent greater than width. With the masthead rig often seen on monohulls, where there is no inner forestay and the pairs of lower shrouds are swept forward and aft very little, the mast's sectional length had best be 40 percent greater than width, as Gerr recommends.

The drawing shows the mast sections for Night Heron and *Dandy.* Weight is even more important a consideration in Heron's mast because the owner will be raising and lowering it frequently for trailering. The *Dandy* mast section must be bigger because the spar is taller and because catamarans are stiffer—they impose greater loads on masts—than trimarans. Having had trouble with slides in the past, I was here interested in incorporating a bolt-rope groove and was willing to put in the forward triangles in order to round the spar and make it more aerodynamic. I now consider this mast deeper fore-and-aft than it needs to be, and when we raise it (which we do only once a year), we use the boom as a gin pole.

The plywood staves of both these masts are $^3/_8$ inch, which is thinner than boards could be, but like ply planking in hulls, ply mast staves can be thinner. Both masts have blocking to reinforce them where the tang bolts go, at the masthead for the main halyard sheave and at the butt for the tabernacle bolt, halyard winches,

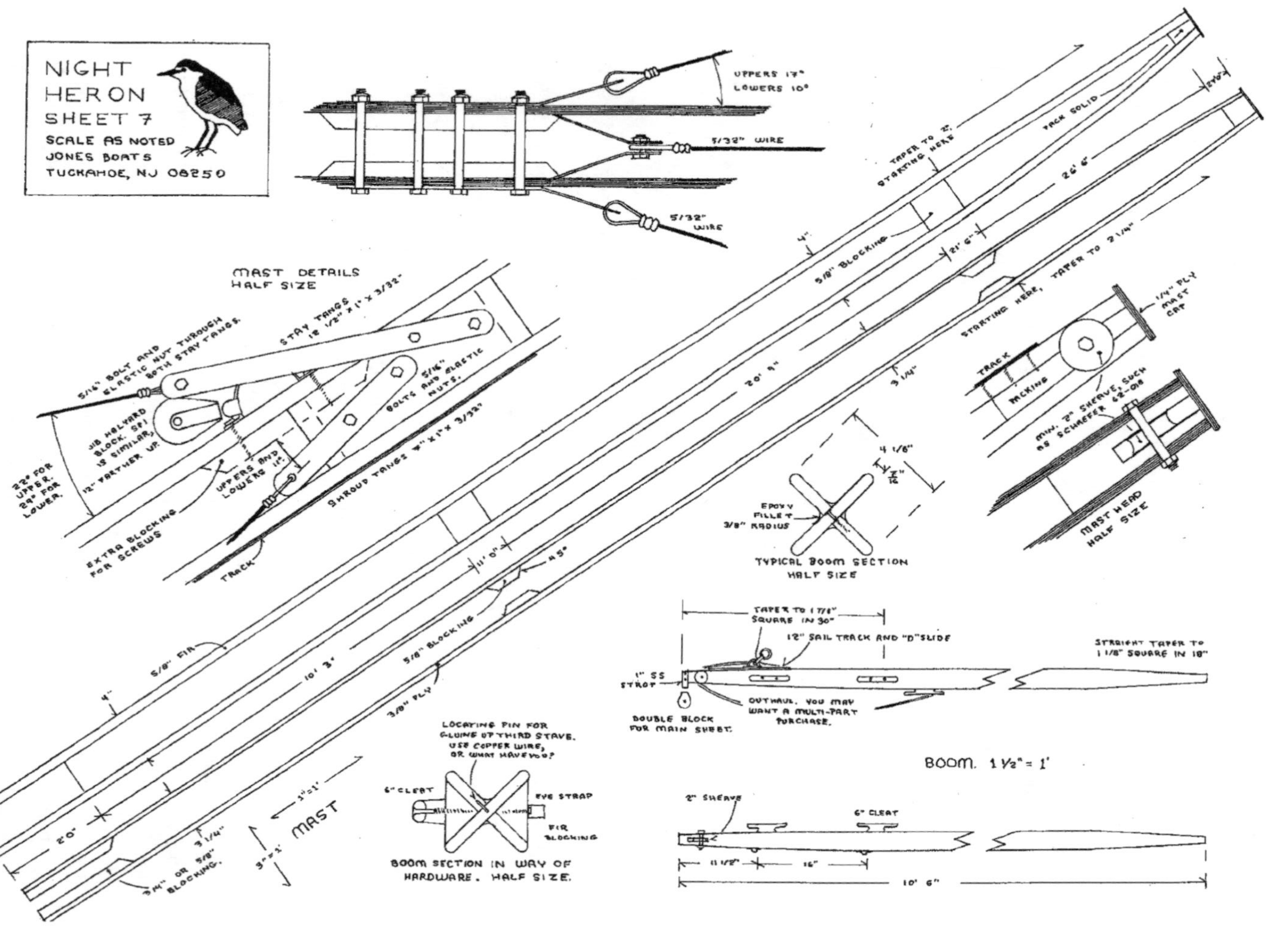


Night Heron spar drawing

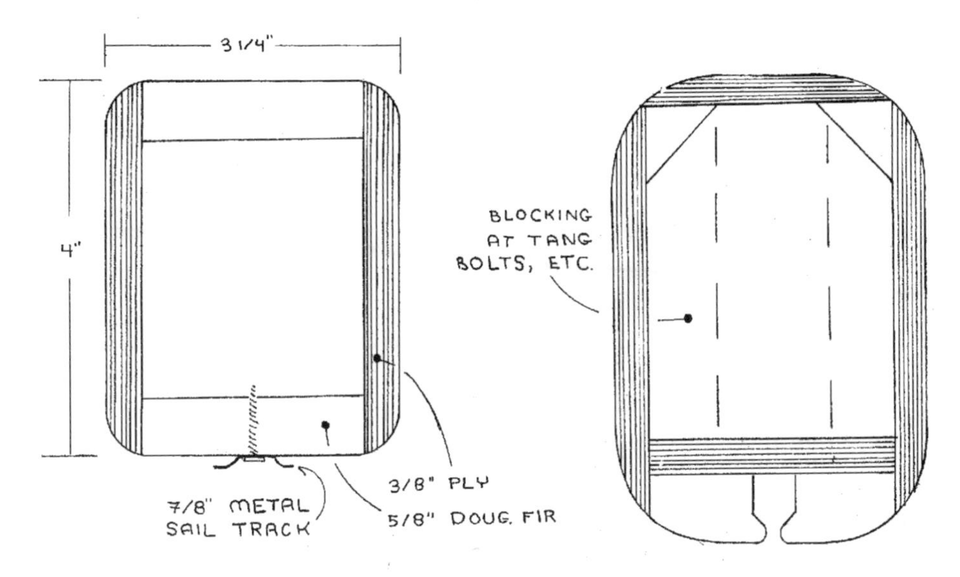


Mast sections of Night Heron (left) and *Dandy* (right)

and cleats. In both, the top 6 or 7 feet is free-standing; that is where all the taper goes, above the blocking and the upper tang bolts.

I started building Night Heron's mast by selecting from my pile of $^3/_4$ -inch Douglas fir (from which the boat's framing was to be made) the lightest-weight and straightest-grain boards. That was dangerous, because spar-making is so interesting that it is tempting to go right on with the mast and leave the hulls until later. That's what I did, in fact, although the ply side staves are more economically cut from leftovers after the boat is built. There is a compensating advantage to this unorthodox schedule: if the strongback for the hulls has been set up, with cross spawls but without frames, it makes an excellent spar bench nearly as long as the mast.

The Heron spar drawing shows the mast running diagonally across the sheet. To show the detail, the scale is three times less reduced in section than in length, which may take some getting used to. The upper drawing shows the mast from the side and the lower one from forward. There are 2 feet of blocking at the butt, 9 inches at the tangs, and 6 inches at masthead.

The fir forward and aft staves were ripped from $^3/_4$ -inch stock and were then thickness-planed to $^5/_8$ inch, which saved nearly 4 pounds. An 8:1 scarf seems safest for spars, though 6:1 certainly is enough for hull stringers and other less critical work. These staves were then glued up with the blocking, bending one to give the forward taper but making sure that the aft one on which the sail track goes remained straight. Some temporary chunks—short pieces of two-by-four ripped to 2 $^3/_4$ inches—were used between the glued areas to help keep the staves parallel.

In all spar-making permanent fastenings should be avoided wherever possible, both for their weight and for the excellent possibility that a tool such as a plane may find them later.

Glue alone was always used, even in the days of not-so-good glue. Then it was recommended to keep moisture out of the spar with varnish, not paint, so that one could peer through the varnish to see how the glue (usually casein) was holding up. Thank God those days are over. Nowadays spars are best painted, like everything else that is exposed to the sun.

The next step was to plane the thwartships taper into both staves simultaneously. Night Heron's mast tapers from 3 $^1/_4$ inches to 2 $^1/_4$ inches, so the staves taper from 2 $^1/_2$ inches to 1 $^1/_2$ inches with an equal amount taken out of each side. They could taper more, but the main halyard sheave is $^5/_8$ inch thick, and it needs some meat on each side. The taper can be the arc of a circle, although even the conservative *Skene's* allows us to take off a bit more wood than that. The taper should not be a straight line because the mast would then be too bendy between the upper shrouds and head.

The plywood side staves were then ripped to 4 inches wide, and a conservative 12:1 scarf was used. The first scarfs were glued on the bench to such length that the final scarfs, which were glued on the mast, fell on the tang blocking, where they could be well clamped. That worked out to two pieces each side about 11 feet long and one 6 feet long. For gluing the ply to the first side, temporary chunks were used as earlier, with the addition of a couple of tapered chunks between the uppers and the masthead. Again, care was taken to keep the aft stave straight.

Unless you have very many clamps in your shop, the ply staves are best held down while the glue sets with small-head nails—say 1 inch by 18 gauge—driven through pieces of $^1/_8$ -inch scrap. The scrap can be peeled off and the nails pulled after the glue sets. Designer Dick Newick and his crew used millions of these scraps while building the cold-molded proa *Cheers* in the days before vacuum bagging. They called them *dufers*, as in "that will dufer now." Jim Wharram points out that another way to get even gluing pressure on a horizontal surface is to cover it with plastic bags filled with water.

The final step in making plywood masts is rounding the corners; since there were no plywood triangles in the Night Heron mast, $^1/_2$ -inch radius was enough. My mast weighed 36 $^1/_2$ pounds. The lightest aluminum extrusion that I'd dare use on Heron would be 3 inches by 4 inches, 27 feet of which weighs 42 pounds. True, paint and sail track on the wooden mast makes up much of the difference, but its center of gravity is far below the aluminum one's. That results from the taper, which also has a slight aerodynamic advantage and a very great esthetic one. The aluminum mast is less trouble to buy than the wooden one is to build; but hey, shipmates, what are we doing out here?

As for running spars, a sprit should be square in section, as should be all spars that are stressed only in compression, such as spinnaker poles. Length times 0.01 is big enough, and the spar should be tapered, especially at the top. Gaff section height should be length times 0.02 and width about half that. A gaff can be tapered to half its height at the ends, with all the taper taken in the top side. Width probably tapers less. The gaff should then be rounded.

A boom is the running spar that nearly all sailboats have, and the sail always sets better if it is loose-footed, although in very big boats having the sail foot on track makes sailhandling easier. Sails are generally not loose footed on production cruising boats because owners like to stow the sail on the boom to avoid the trouble of taking it off the spar. (The El Toro rule book says that only sails made before 1956 can have a loose foot. In other words, forty-five years ago someone had figured out that a loose-footed sail made a better Toro.)

You may have noticed that Al Whitehead's Bobcat, described in chapter 2, has a sprit boom, which saves the cost of a gooseneck and in theory vangs the sail. However, for ideal shape a sail should be fuller downwind than up, and if the tension on a sprit boom is eased, the vanging becomes less effective. A sprit boom that works well needs two attachments at the mast, one to hang from and the other to adjust foot tension. Otherwise, trying to ease foot tension only results in lowering the boom, especially in light air. And on one tack a sprit boom creases the sail much more deeply than a vertical sprit, which doesn't improve sail shape.

Nevertheless sprit booms are attractive, and to remedy their faults a builder sometimes goes to the trouble of making a laminated wishbone boom, which has a loop forward and extends all the way around both sides of the sail. Sprit booms can be awkward to handle when reefing, but wishbone booms can be even more so. In the Canary Islands we once met an older Belgian couple aboard their Freedom 40, which sported not only wishbone booms but also sails sleeved over the masts. They told us that at anchor they reefed without difficulty, but once a gale arose quickly while they were under sail, near Malta. They didn't dare reef and so ran off for several days under full sail, fetching up eventually in Greek waters.

A sprit boom should be square with a section 0.0125 times length, and it should stick out at least a foot past the mast to allow a rope to go from its forward end around the mast and forward again to a cleat. Once I tried to make a perfect sprit boom that would not crease the sail on either tack. In fact, I made two of them

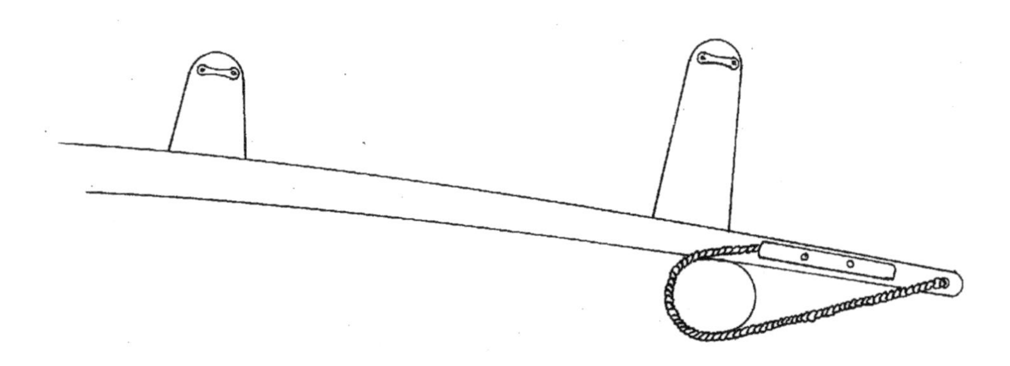

Forward end of curved, balanced sprit boom

because they were to be used on *Dandy*'s first rig, which had a mast on each hull. The sad result of that I have related in *Multihull Voyaging*.

The drawing shows one of these booms, laminated to 6 inches of arc in 12 feet, 3 inches of length. The clew end had a spinnaker pole fitting that was very easy to clip onto and release from the sail. The forward end had two $^1/_2$ -inch plywood ears to which the rope that hung the boom from the mast was attached, the forward one for the full sail and the aft one for reefed. The exact balance point that allowed the boom to be horizontal and not interfere with the sail was arrived at by experiment, not by mathematics; and it did work perfectly in a variety of conditions. It wasn't these booms that doomed the biplane rig.

Just as Phil Bolger likes to say that the Birdwatcher sailboat type is his one original contribution to yacht design, I like to imagine that the X-section spar is mine, although probably someone else thought of it first. The original Lightning boom, designed by Sparkman and Stephens in 1938, is a T-section, but an X-section offers even greater stiffness and less weight. The netting beam of *Dandy* and the masts of our El Toros are X-sections, although I wouldn't recommend such a section for a Bermudan rig that was closer-winded (Toros tack in about 110 degrees). But for the boom of a loose-footed mainsail, an X-section can't be beat.

Such a boom should have a section that is 0.022 times length. At the bottom of the Night Heron spar drawing is the 3-inch-square boom, with staves the same thickness they would be in a box spar, $^7/_{16}$ inch. The weight saved over the box spar is about 25 percent, and I believe the X-section is stiffer. It is easier to make and easier to add blocking to for hardware fastenings—and if hardware needs to be moved or added, there's little work to chopping out one V-shaped block and putting in another, whereas with a hollow box spar extra blocking can be added only to the outside.

This boom is made by ripping an 11-foot-by-4 $^1/_8$ -inch-by-1-inch fir board edgeways and then ripping one piece flatways to get two staves 1 $^{27}/_{32}$ inches wide. One of these is then glued and screwed to the 4 $^1/_8$ -inch stave (in this case, I do leave the fastenings in). The last stave can be glued and clamped but not screwed. To help center it small holes are drilled in each piece about a foot on center, and bronze or copper nails with their heads clipped off are inserted, like locating dowels. An epoxy fillet in each corner then strengthens the joins. Dave Carnell claims—and he usually has test results to back up his claims—that a fillet with a $^1/_4$ -inch radius is the biggest size that adds any strength, but I feel more comfortable with a fillet radius of $^1/_8$ inch for each inch of spar dimension, which in this case is $^3/_8$ inch. I can't justify that figure, however.

The X-boom is, of course, easy as pie to taper; it's more convenient to round the corners before putting in the hardware blocking. That's really all there is to it, and I can't figure out why everyone doesn't have an X-section boom.

Like the forward end of a sprit boom, the aft end of a loose-footed boom should have one attachment to hold the boom up and another to adjust foot tension. With very small solid booms, as on the mini-garvey, a light line goes from the sail clew around the boom, while another goes from the sail clew to a hole in the boom end. For bigger sails the easiest solution is to attach a piece of sail track about 9 inches long to the top of the boom and put a slide on the track. Then the sail is attached to the slide with a snapshackle, and a line from the aft end of the slide goes through a sheave on the boom end to a cleat underneath the boom. In a rising wind, increasing foot tension without luffing the sail requires considerable force, and big boats often have a winch for that purpose. On *Dandy,* whose sail foot extends 11 feet, 3 inches, we find a three-part purchase is sufficient.

# Structural Plexiglas

Many years ago when I was working to complete a 51-foot trimaran, the owner finagled some 1-inch Plexiglas that was being discarded by a horse-racing track. It had been used to separate the cashiers from the public, and it was literally bulletproof. Handling it, I couldn't help thinking that it had structural possibilities, but we used it only for portlights (which was a bad idea, on a weight-sensitive multihull). Sadly, the owner made me lay it out in such a way that all the lettering saying "please count change before leaving window" was cut to waste.

Bill Powers recently brought me a 20-foot lobsterboat hull that he had purchased in Maine. It was molded glass with a glass-over-ply deck and a huge outboard. He wanted me to build seats and a console and generally make the boat ready for the water. For furniture he wanted epoxy glass over $^1/_2$-inch marine fir plywood. I suggested using $^1/_2$-inch painted Plexiglas instead of ply for the lifting panels of the seats, and he agreed.

Plexiglas weighs more than water and about twice as much as glassed fir ply. Half-inch Plexiglas costs about $150 a sheet, or three times the price of ply. It may not be as stiff or strong, so it is hardly the ideal material for every part of a boat. But if plywood is to be glassed, wrapping the glass around any smaller parts and small-radius corners is tedious work and not likely to produce a tidy finish. For example, try glassing around a finger hole in a lifting panel and then survey the result. If your time is worth anything, the Plexiglas alternative starts to seem very inexpensive: A hole saw cuts the finger hole smoothly. The edges of the hole are rounded with sandpaper wrapped around a dowel, and the flat surfaces are sanded until cloudy to give the primer paint a better grip.

To preserve the unscratched surface, Plexiglas comes with a protective wrapper stuck to each side, sometimes paper, sometimes plastic. The paper is better both because it doesn't pucker up as the plastic often does and because cutting lines can be drawn on it with a pencil. Cutting lines can be drawn on plastic wrapper less satisfactorily with a felt-tipped pen. If the paper wrapper has been on the Plexiglas a year or more, it is reluctant to come off, but rubbing alcohol dissolves the hardened stickum; it doesn't work quickly, but it does work.

Almost any tool that cuts wood cuts Plexiglas, but the tool must be sharp; the Plexiglas also dulls it more quickly than wood does. All cuts must be made in one

Plexiglas and teak deck box

pass, or else the tool must be backed out of the cut while it's still running. Often when cutting a sheet of plywood—with a Skilsaw, for example—I release the trigger and walk partway around the sheet to be in a better position to complete the cut. Try this step with Plexiglas and the heat of the blade may fuse it to the plastic, and when you turn on the saw again the glass may shatter or crack. Drill bits can do that, too.

The worst electric saw for cutting Plexiglas is a saber saw because a tiny blade area does all the cutting and accumulates all the heat. The best is a bandsaw because much of the blade's heat is transferred to the wheels with each revolution. But sharp circular saws can also be used, although I have found that my Skilsaw takes out fair-sized chunks and throws them in my face, so a face shield is recommended. Table saws are less hazardous, as most chips and dust go down to the ground. My Plexiglas supplier says that the more teeth the blade has, the better; he uses a plywood-cutting blade in his panel-cutting saw.

A short time ago a cockpit step-box was stolen from Bill Powers's other boat, a 40-foot sedan cruiser, and he asked me to make him a new one. I went down to the dealer to measure and study the kind of box he'd lost. It had dimensions of 24 inches by 16 inches by 13 inches (the last half the height from cockpit sole to rail) and was built in a mold with an odd mix of fiberglass and mediocre plywood. In addition to being a step, its hinged lid opened, making the box a handy place for stowing docking lines. Not having a mold, I told Bill I couldn't duplicate the box. I could build him one out of teak boards, but teak doesn't glue well and I couldn't guarantee how long it would last, even with a bunch of framing inside to receive metal fasteners. And given the price of teak, it would have to be more expensive than the factory box.

Certainly such a box could have been made from plywood, and I did have on hand enough $^1/_2$ -inch sapele scrap; but Bill mistrusts unglassed ply, and although I don't agree with him, he's paying for it. Glassing a box like this would make glassing lift-out seat panels look like child's play, so I suggested that it be built of Plexiglas with a teak top and handles (Bill likes teak). He went for it, so now I wasn't just cutting out Plexiglas shapes; I was joining them, too. Epoxy glues sanded Plexiglas moderately well, but it's better to use a fillet on all inside corners because the glue can't grab any pores the way it can in wood. To compensate, the fillet makes the glue join larger.

Mechanical fastenings also were used in this box to align the panels and also to give whatever extra strength they could. Even the sharpest drill bit tends to pull through the last morsel of stock once the drill tip breaks through into air. In wood that just makes a few feathers on the bottom side, easily sanded off. In Plexiglas the panel tends to leap up the drill flutes toward the chuck, and when it gets to the end of the flutes or to the chuck it can crack. Whenever possible I drill Plexiglas in the drillpress, putting great weight on the panel or even clamping it to the table and drilling the last bit of the hole very slowly. When drilling Plexiglas portlights for a wooden boat, holes should be oversize to allow for different expansion and contraction rates; but when working with a solid Plexiglas structure oversizing isn't necessary.

Flat-headed fastenings were used, but they were driven carefully with a hand screwdriver, not a battery-powered one, and taken up only enough to close the join. Wood screws are no good in Plexiglas because the plastic does not yield to accommodate the screw threads, as wood fibers do. Machine screws work better. I used a pilot hole only $^1/_{32}$ inch smaller than the clearance hole and ran a tap into it.

All these steps and considerations make building with Plexiglas sound like a considerable nuisance, and you may ask, Why bother? One reason is that Plexiglas is absolutely not biodegradable provided paint is kept on it, and the Plexiglas box or seat or shelf or console that you build today will still be around when the Parthenon has crumbled to dust. (That may or may not be a good thing.) Another is that Plexiglas objects have a finish more perfect than any other material I've worked with. You cannot, with wood or fiberglass or metal, dare hope for the perfection that comes with Plexiglas.

This box was painted with one-pot polyurethane. To hold the lid open it has one of those bend-to-collapse stainless steel springs. Sitting on its rubber feet, which do not mar the cockpit sole, it looks so perfect that I'm astonished I made it myself.

## Hardware without Welding

In some ways, metal is more satisfying to work with than wood. It's even more dimensionally stable than plywood, so the object you make today will be the same size and shape ten years from now, and the joins won't have opened. It has no imperfections such as irregular grain, sapwood, checks, or sap pockets, so it holds no unpleasant surprises, and it fastens reliably and cuts predictably with saw, shears, or drill. Torch cutting and fastening are another matter.

Alas! Carol and I signed up for a welding course at the local vocational technical school but discovered that when we put on our welding masks we couldn't see through them. That's not uncommon. In Dakar, Senegal, a motel handyman welded up mild steel gudgeons and pintles for our trimaran because the stainless steel ones had cracked and been repaired so often—stainless fatigues from vibration. He would align the mild steel pieces, hold the electric torch right above them, close his eyes, and zap them. Then he would open his eyes and giggle. His gudgeons and pintles did get us home.

A mask used for brazing is less opaque, and Carol spent the rest of the welding class brazing up an angle-iron frame to hold the certificate she would get when she graduated. I explained my boatbuilding business to the teacher and asked whether brazing would make a stronger join than the propane soldering I was then doing. He said it really wouldn't and would require buying and keeping serviceable all the same equipment that welding required. Silver solder joins were a bit stronger than tin, but the solder was ruinously expensive. When I really need welding or when I need $^1/_4$-inch stainless steel cut, I take it down to Yank's, and eventually the yard does it. For all steel Yank now uses a "plasma cutter," which is said to go through stainless as if it were cheese and requires that you wear a mask no darker than you would for brazing.

A circular saw or table saw can be used to cut stainless steel. An old plywood-cutting blade is best. It starts working its way through the material when it gets red hot; the heat is what does the work, not the teeth, and it continues to work long after the teeth are worn off. It's a bloody business.

Stainless can also be cut with a hacksaw and patience. The intermediate blade with 24 teeth per inch is best for nearly all work. For very thin sheet metal a 32-tooth-per-inch blade is recommended, but very thin sheet metal can usually be cut with shears. Two things are important when cutting metal whether you use a saw or drill: run the tool as slowly as you can, and keep the work lubricated. Hacksawing stainless steel is slow work, and it often seems that more strokes per minute gets through faster, as it does with wood. But each tooth of a saw needs to get a bite on metal, and if you watch work in a machine shop you will notice that the electric tools used for metal turn more slowly than woodworking tools.

Lubrication lengthens the tool's life and also speeds the cutting. The time to add another drop of oil is when the last one you put on smokes, which even a manual hacksaw moving slowly in stainless steel does. Cutting oil is made especially for this purpose, and no doubt machine shops use it, but regular engine oil is a good deal better than nothing.

For joining metal parts without welding, bolts are the usual choice; the nuts should be elastic stop-nuts. Ordinary nuts usually work with wood, which is itself elastic and so puts pressure on the nuts to keep them from unwinding, but metal doesn't do that. Sometimes there won't be space for a nut in the part you're making; then it is necessary to drill a hole and tap threads into it. It's good to have a half-dozen taps on hand, ranging in size from 10-24 to $^1/_2$ inch. Having a half-dozen dies that can refresh the threads on bolts allows you to cut down a longer bolt, so you need not have every length of every imaginable bolt on hand in the shop. A nut does not do the work of a die, even when put on before the cutting and wound off afterward. As with all cutting, stainless steel wears out taps and dies quickly.

In fact, stainless steel is a plaguey material, nearly as annoying as oak, and bronze is every way pleasanter to work with, and is adequately strong and less likely to fatigue. But stainless is what's around and at ever more attractive prices. You may have noticed that stainless sheet-metal screws, now commonly used for wood as well, have become cheap enough to drive galvanized screws off the market. Buying bronze stock—say 1 $^{1}/_{4}$ inches by $^{1}/_{8}$ inch for chainplates—requires immense research, and when you find it you may need to buy a wholesale quantity. But the dreadful stainless is ubiquitous.

In making metal parts that will be above the waterline, I don't bother much with the galvanic series. Twelve years ago I built a kick-up rudder for our trimaran using aluminum plates and stainless steel bolts. The present owner of the boat has not reported problems, although rudder fittings are wet with salt water whenever a boat is under way, and aluminum and stainless are fairly far apart in the series. However, bronze is a good deal farther from aluminum, and that is why on aged Sunfish the bronze gooseneck has often destroyed the aluminum boom. Bedding would prevent that process or at least slow it down, but the makers proved slow learners.

Below waterline the galvanic series requires the closest attention, and maybe wood should appear in it as well as metal. We once saw a beautiful forty-year-old sloop at a yard in Virginia. The ribs had been diagonally strapped with bronze and the carvel mahogany planking notched over it. The yard owner said that the bronze had eaten up the mahogany. George Harris, who worked as a marine surveyor all his life, heard this story and said that *any* two materials that touched and were constantly in salt water were trouble, for one would eat the other. Think of that and burst out crying!

Two examples may show the ease and pleasure with which boat hardware can be made of metal. The first is a copper ventilator made from 3-inch plumbing pipe. Two pieces are needed, one 2 inches long and the other 4 $^{1}/_{2}$ inches long. The pipe is marked with a felt-tipped pen, and a piece of typing paper is wrapped tightly around it as a marking guide. Pipe can be cut on a bandsaw using a blade that is past its prime for cutting wood (brass and aluminum can also be cut on a bandsaw, dulling the blade only a little). More laboriously, the copper (as well as brass and aluminum) can be cut with a hacksaw, of course. The 4 $^{1}/_{2}$ -inch piece is then cut longitudinally to yield two unequal pieces, one with 4 $^{1}/_{2}$ inches and the other with 5 inches of the total circumference.

The smaller piece is hammered flat to make the base plate. It is best to clamp a two-by-four on edge and tap gently with a hammer. A 3-inch hole can then be cut through it with a saber saw using a hacksaw blade. Saber saws don't always go where they're supposed to, but if the hole is somewhat irregular the solder will bridge it later. Solder bridges just like epoxy.

A similar hole is cut in a piece of thick hard wood such as oak and blocked up on a bench so that when the 2-inch length of pipe is dropped into it, about $^{3}/_{16}$ inch

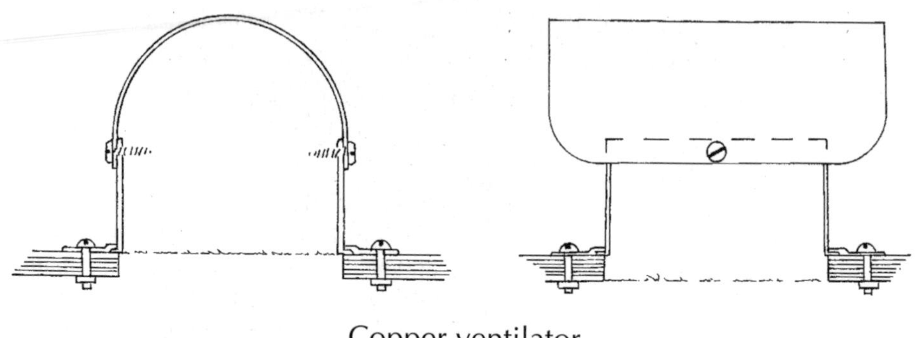

Copper ventilator

protrudes. As when flattening the base plate, bending the flange over is best done with light taps, a little at a time. Keep tapping, and keep turning the pipe in the oak hole. Copper likes to be worked that way and resents sharp commands. The bottom of the flange is not likely to be perfectly flat but can be improved on the flat side of a grinding wheel, holding its center over the arbor, or more slowly on a piece of emery cloth on a flat surface.

Like glue but even more so, solder grabs best on a clean surface, and like all metals, copper oxidizes, so it needs to be sanded bright before soldering. The sheet-metal screws through the top into the vertical tube, which can be seen in the drawing, help to hold the parts in position for soldering and strengthen the join for the vent's whole life, in case somebody trips over it or a rope gets caught in it. Tin solder is not an especially strong material. The base plate, if not perfectly flat, may need small clamps to pull it down to the flange until the solder cools.

A vent like this never takes in rainwater if the boat is on anchor and can swing to the wind. Even at a dock or on dry land, powerful wind and rain bring in only a few drops, and because the space is ventilated the drops soon dry. The space is drier than it would be from condensation if it were unvented. However, when we start a long ocean passage where we expect green seas to be breaking over the boat, we plug a vent like this with a bit of sponge threaded through one side of the vent and out the other. We try to remember to take the sponge out when we reach port.

The bit of bronze screening shown in the drawing, which is stuck into the bedding compound on deck before the vent is bolted down (it can also be tacked up underneath later) is to keep wasps out. They like wooden boats, at least in New Jersey, and as we leave hatches cracked to add to ventilation, we have bronze screens for them, too, which we leave at home when going on a voyage. Don't bother with plastic screening: wasps eat it right up (or perhaps use it to reinforce their nests).

The second metal-working example (although these two do not exhaust what you can do with metal, even if you're welding-impaired) is a gooseneck for a 25- to 35-foot sailboat. There used to be several sources for goosenecks like this one, meant for wooden spars, but aluminum is such a common spar material nowadays that I don't know where you'd buy a gooseneck for wood. The manufactured ones were

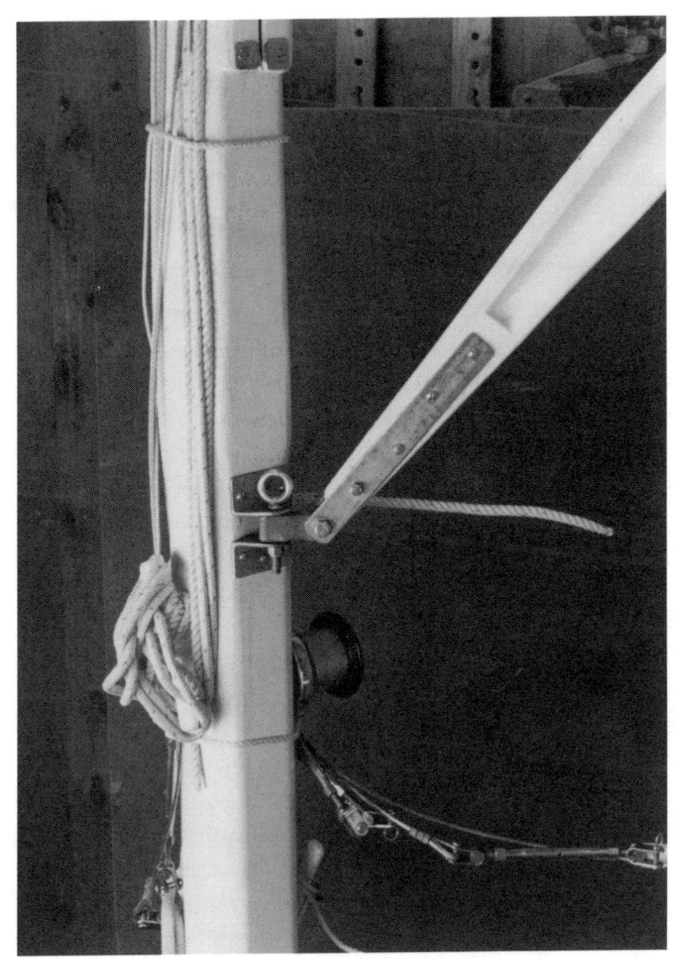

*Dandy's* handmade gooseneck

bronze castings and very pretty. If you come upon a foundryman who will cast from your patterns, you can make the parts out of wood, but most foundrymen prefer to work from drawings and make the patterns themselves, because not all shapes allow the metal to be poured cleanly.

The photograph shows *Dandy's* gooseneck, which has only five shop-made parts, a few $^{5}/_{16}$ -inch bolts, and some wood screws. The photo also shows *Dandy's* X-section boom and plywood mast. Screwed to the mast are two pieces of stainless angle iron 2 $^{3}/_{4}$ inches long by 1 $^{1}/_{2}$ inches square by $^{1}/_{8}$ inch thick. I did take the trouble to cut the corners off at 45 degrees, although it was tedious. The boom straps are 1 $^{1}/_{4}$ inches by $^{1}/_{8}$ inch, and the toggle is a piece of solid brass l inch square and 3 inches long. Brass wears away faster than bronze or stainless, but only brass was available, and the gooseneck is okay after seven seasons.

The eyebolt for attaching the mainsail tack is boughten. The bottom end was not cut short and has proven useful to take a rope around when tying in a reef. To make such a part you would need an abundance of tools and skills. Recently we saw a 31-foot ketch whose lines had evolved from an H-28. Both hull and deck were cold molded of mahogany veneers, and the owner had made every single metal part himself of aluminum bronze, which is not a common alloy but which he assured us is the best. To make the turnbuckles he had to have left-handed as well as right-handed taps and dies, of course, and if he followed Herreshoff's rigging schedule, there would have been three different sizes of turnbuckle, but I failed to notice that. With rudder and rudder hardware, aluminum bronze anchor and anchor chain, and so on, he had been working on the boat for twenty-five years, and it still wasn't finished. Clearly the building was the entertainment for him, not the sailing. But even if you enjoy both building and sailing, there is a good deal of metal hardware on a boat that you can make for yourself, despite limited tools and skills. It will work well and last well, and it will be a pleasure.

# Cleats

The only plan for making cleats that I know of is in a very old boatbuilding book by Edson I. Schock, son of the great powerboat designer Edson B. Schock. Schock's cleats are elegant, but I like my scantlings a bit more robust. Wood is not as strong as metal, so wood cleats must naturally be meatier than metal cleats, but even the heaviest oak weighs only 30 percent of what aluminum does, so wood cleats can afford to have heavier scantlings.

However, I don't recommend oak for cleat making or for any other bits and pieces if you can find iroko. Iriko is an African hardwood often found in European lumberyards, and builders used to oak find iroko miraculously stable—almost as unchanging in different temperatures and humidities as metal. On John Guidera's Melonseed I tried for an elegant finish by filling the screw heads of the well-seasoned oak cleats with thickened epoxy, but two years later the oak had shrunk until the epoxy stood proud $^1/_8$ inch. It wouldn't have happened to iroko, and the fine grain takes paint well and never splits. Tropical woods are usually easier to finish because the tree grows year round, so the wood doesn't have alternating rings of hard winter and soft summer wood. Teak does have alternating rings but is hard and dimensionally stable enough to make good cleats. The silica in it dulls tools quickly though, and its oils make a poor surface for paint, varnish, or glue, although a preliminary wipe-down with acetone helps. Chapelle says that locust also makes good cleats.

*Skene's* says that cleats should be 1 inch long for every $^1/_{16}$ -inch line diameter. For $^3/_8$ -inch line, therefore, a cleat should measure 6 inches long, but I prefer 7 inches, and my drawing is based on that figure. The numbers shown are all multi-

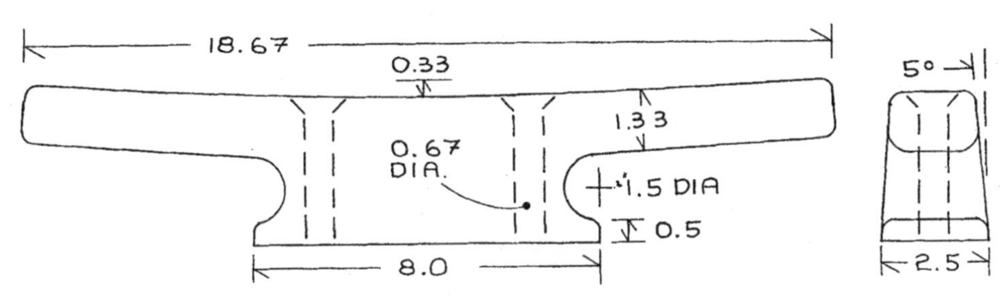


Wooden cleat; numbers are multipliers of rope diameter

pliers, so for $^3/_8$-inch line the base is 8 times $^3/_8$ inch, or 3 inches long; the fastenings are 0.67 times $^3/_8$, or $^1/_4$ inch; and so on. Larger or smaller line can use the same multipliers. However, a mooring cleat should be bigger because we all tend to take more turns and hitches on it before leaving our boat to its fate than we do with other lines. Ten inches is a good cleat size for $^3/_8$-inch anchor rode, and other dimensions should scale up in proportion.

Usually when we need a cleat we need more than one of them, and they are best made in batches of four or five. Seven-inch cleats are laid out on a board 1 inch by 1 $^3/_8$ inches, end-to-end, with perhaps a slight gap between them. Lay the board down on the 1 $^3/_8$-inch side and make all the cuts first, the hole preferably with a drillpress and the others with a bandsaw. Then bevel the sides on the table saw and smooth them off on the joiner. If you don't have these power tools, all these steps can be performed with hand tools, but the work will go slower.

Now is the time to drill the fastening holes and countersink them, because if you round over first and round over too much, the heads of the fastenings may stand proud of the wood, where they will harass the rope. Marks for fastening holes are brought up from the base, on which you marked the cleat centerlines before making any other marks, because by this time you have cut away all the other original surfaces. When rough rounding the top edges you can use a 7-inch grinder with 60-grit disc, one end of the board in the vise and the other end easy to get at. I say board, but at this point the work looks more like those strings of stick figures holding hands that children like to cut out of folded paper and open like an accordion.

Notice that by waiting to cut the cleats to 7-inch lengths your fingers (and perhaps other parts) don't have to get near the cutting edges of the tools. I had a young friend who wanted to cut a very small piece of wood with a Skilsaw, the only power saw on hand. He wired the trigger on and the guard back, sat down in a chair, put the saw between his knees, and plugged her in. When he put the teeny piece of wood into the blade the saw jumped at him, and he was rushed to the hospital. He needed quite a few stitches, but the doctor said he was lucky and would still be able to have a family some day.

At this point you must remark the spacing between the cleats and cut them apart. A cut at right angles to the curved cleat top makes prettier work than a cut at right angles to the base. Now the lower edges of the horns and base must be rounded, and again I rough them out with a disc grinder, clamping first one horn and then the other in the vise. For rounding the semicircle between each horn and base I use a rat-tail file. It is easy to round the horn ends into pretty semicircles, but the more the ends are rounded, the fewer the wraps of rope that can be made. Rounding the ends overmuch in effect makes the cleat shorter. That is sometimes done with metal cleats and always done with plastic ones, for what reason I don't know.

After you have rough-sanded the cleats, they need only to be finish-sanded. I leave the particulars to your imagination and elbow grease, but note that in finishing the semicircle, wrapping sandpaper around a piece of dowel somewhat smaller than the semicircle is preferable to wrapping it around your finger.

These cleats can secure 90 percent of the lines that need holding on a boat. For docking lines, halyards, or sheets that are behind a winch or purchase, nothing is better. Hard paint on them, such as polyurethane, takes the abrasion better than latex. But on sailboats a few situations need specialty cleats, which are best bought ready-made. No sane sailor would think of mounting a wooden cleat on the leech of a sail to take the leech line, for example. And on a light daysailer such as the mini-garvey, a gust of wind demands that the sheet be let go quickly. By the time a hitch is taken off a horned cleat and the wraps unwound, the mast and the skipper may well be in the water. So on the mini-garvey we use a swiveling bull's-eye cam cleat, which can be let go instantly, mounted on the aft end of the board trunk. Such cleats cost $30 or more, but because I make our other cleats of wood I have the money in hand.

# Tools

Among both amateur and professional craftsmen there are some who are more interested in the tools than the product the tools make. That's okay, but it's not my point of view, and it's worth remembering that most boatbuilding jobs can be done without any power tools at all. Recently *WoodenBoat* magazine ran a three-part article about building a flatiron with hand tools alone. Carol and I have never had access to power tools on the long voyages we've taken, and one memorable summer we ripped the bottom out of our trimaran on a lava reef and six weeks later broke her mast in half. The bottom was replaced and a new mast was made entirely with hand tools, although commercial shops did the ripping for us.

In the old days the simplest boats were built with hand tools—and very few tools at that. A skiff made with boards and fastened with nails called for little more than saw, plane, hammer, and paint brush. Today the simplest boats are likely to be taped seam, and who would start on one of them without having at least an electric sander?

When you buy a tool today, what purports to be an instruction book turns out to be a list of WARNINGS, always printed in capitals, about the dangers the tool has in store for you. These "warnings" are written by attorneys to protect the manufacturers from lawsuits. However, all tools, including hand tools, are dangerous to some extent. I lost part of a finger in a plane, and I know very few longtime woodworkers who haven't lost fingers or hurt themselves severely in other ways. The most dangerous time comes when you are working with a new tool; you don't know exactly how the tool can do damage. With most new tools I've bought I've damaged either myself or the tool in the first couple of uses, often in ways not covered in the warnings, so it pays to think about a tool before using it. Some tools, such as a radial-arm saw (which sucks the work toward the blade) or a shaper are just so inherently dangerous that I won't go near them. Any electric tool that turns fast, such as a plane or router or grinder, does more damage faster than a slower-turning tool, such as an electric drill.

Like the materials we use, tools are not as good as they used to be. In a stable economy such as feudalism or communism there is no incentive to cheapen a product, but in capitalism the money saved can go into lowering the price to attract more customers or into the manufacturer's pocket. Forty years ago I bought from Sears their least expensive electric drill. It cost $15, and it still runs today, although the bearings do gargle a bit. In the interval inflation has been about fivefold, and the least expensive drill today sells for about $20. It won't be running forty years from now.

The crude tools that are used for framing houses are not ideal for building the lightweight boats described in this book. When I was a house carpenter I had a 20-ounce hammer for framing and a 13-ounce hammer for finish work. Now I only use the 20 when I can't find the 13.

Saws are a more complex subject. If you are left-handed like me you probably prefer a worm-drive Skilsaw where the armature is parallel to the blade face. If you are right-handed you probably prefer the more conventional Skilsaw with armature at a right angle to the blade. I don't know why that's so. In either case you make more accurate cuts with only one hand on the saw. When you start a cut, your other hand is needed to hold the guard up, but after that the second hand only tries to push the saw sideways when it wanders from the line. But a saw should be steered, not pushed sideways. Cutting with one hand only is also better for bandsaw work and for freehand work on the table saw.

For round holes to 3 inches in diameter a holesaw does the neatest job, although a set of holesaws can be expensive. Fiberglass or aluminum dulls them quickly, although their steel is too hard for resharpening. Other saws that cut holes in the center of the work don't do so very accurately, except a very big, heavy, power scroll saw, but few of us have enough use or space for such a saw. In a small shop every stationary power tool should be light enough to be moved around to suit the job at hand. Lightweight scroll saws with sheet-metal tables want to grab the work and jump up and down with it instead of cutting the wood, unless they're clamped to a bench. With an electric or hand jigsaw, the problems are keeping the cut square to the face of the work and splintering wood. For squareness, one-hand operation again helps, and if you have a power jigsaw, the table had best not have an angle adjustment. A good one has blade guides just above the table, which also helps keep the cuts square. All wood-cutting blades for these saws are too coarse and tend to splinter the wood, especially plywood. A hacksaw blade cuts more slowly but more smoothly. For cutting fiberglass a jigsaw with carborundum blade cuts cleanly and does not spray dust the way a circular saw does.

Many table saws have fences that do not adequately clamp them on the far side of the table, so the cut wanders according to the sideways pressure that is used to hold the work against the fence. Often such fences have adjustments and can be tuned up. My table saw is small and is surrounded by a plywood extension. The fence is Formica-covered wood with a wooden clamp at each end, so the far end stays put.

For accurate rips a comb is helpful. It is made of springy stable hardwood such as walnut, perhaps 1 inch by 2 inches by 18 inches. One end is cut at 45 degrees, and a series of parallel rips is made in it 6 inches long and $^1/_4$ inch apart. With two clamps attaching it to the saw table, the fingers or tines of such a comb hold the stock against the table saw fence more evenly than hand pressure possibly can.

For a stationary power tool that wood is passed across, whether saw or planer or joiner, a cast-iron table is wonderful; the wood slips smoothly over it. With aluminum or sheet-metal tables the slickness is missing, and the wood tends to move in fits and starts or even wander. Such tables can be improved with wax, for instance by rubbing a candle over them. Other lubricants such as oil may be slicker, but they penetrate the stock deeply and make finishing difficult.

I often use steel table-saw blades rather than carbide ones because I can sharpen them myself at the vise with a file, and that saves the time and mileage of two round-trips to the sharpening service. I also sharpen handsaws, which is quick and easy to do as long as they are not finer than 12 point. Books and pamphlets to explain the sharpening and setting of saws are everywhere available, and the investment in tools and jigs is less than $50.

The hardest tools to sharpen are drill bits, and although some people can do it by eye, I use a "drill-grinding attachment" that is bolted down to the bench beside the grind wheel. The drill bit is put in and the tool is adjusted for diameter and length. Then it is swiveled against the side of the spinning grind-wheel once or twice, and the bit is turned over and again swiveled against the wheel. The resulting flutes are correctly tapered and miraculously symmetrical. The literature says that this attachment only sharpens bits $^1/_8$ inch or larger, but I find that it can be coaxed into sharpening a $^7/_{64}$ -inch bit. The three smallest sizes must still be sharpened by guess and by gosh. These small bits are not expensive and are as easy to break as they are hard to sharpen, so it's best to keep a good number on hand.

Like the sharpening of carbide blades, the sharpening of the long knives of joiners and thickness planers calls for tools and jigs that most of us don't have, and the work is best sent out. For sharpening other flat-bladed tools—chisels, plane blades, and so on—the best tool is a belt sander with fine-grit belt. Even the hardened steel of the blades of an electric plane sharpen quickly on the sander, and because the belt takes the heat away to the rollers, the corners of the blade do not overheat and turn brown, as too often happens on a grind-wheel. These days I use a whetstone for little more than wiping off the wire edge that appears on the back side of a blade to give notice that a new edge has indeed been made.

A belt sander does not make the superfine tool edge that is wanted by cabinetmakers, but different kinds of craftsmanship call for different grades of perfection, and perhaps they also suit the temperament of different craftsmen. Cabinet and furniture making is very fine work and too picky for many of us. Housebuilding is relatively crude (although docks are cruder), and the emphasis is on getting the job done in the shortest time. Boatbuilding is somewhere in between, or should be, and in addition to your fascination with the water and being out on it, you may find that boatbuilding suits your inherent patience, your coordination, and your eyesight. We have all seen boats that are built with the perfection of furniture, and if the building is done for a hobby that may be acceptable, although the environment in which a boat is used spoils that perfection more quickly than the more protected environment in which furniture is usually found. We have also seen boats built to the standard of a framing carpenter, which is visually disappointing and often not watertight. The worst combination is a framing carpenter's building technique combined with a cabinetmaker's finish. Unfortunately we have all seen boats like that, too. In fact, most production fiberglass boats seem to be made that way.

For the building of plywood boats—the majority of the designs in this book—the best fasteners are screws, not nails. Nails don't pull in as strongly, even with a back-up behind them, and as years go by they often work their way out, even if annular, and the heads stand proud of the surface. Slotted screws are best avoided, as the driver blade too readily wanders out of the head and chews up the wood. Various types of screw head hold the driver centered in the screw: Frearson, Phillips (they are *not* the same thing), square-hole, and no doubt others. It is best to make a decision about which type to use as early as possible and to stick to it. I still have two screw-head systems in my shop: in seldom-used sizes the slot-heads I formerly used and in frequently used sizes the Frearson that I am now buying. That often requires changing tools or tool bits in the middle of a glue-up, with begloved hands that are liberally coated with epoxy. Friends have suggested switching to square-hole, but I am resisting the change.

Four types of hole are required for every wood screw: the pilot hole, the clearance hole, the countersink, and the hole above the countersink that is filled with putty or a plug. Drill bits that make the four holes in one shot are well worth their high price, even though they're not the best steel and the glue in plywood dulls them quick. I drill with a 110-volt drill and drive screws with a battery-powered one because it is less powerful, turns more slowly, and has an adjustable slip clutch and so can be prevented from stripping the screw. Battery-powered tools are all the rage now, but the batteries don't last long and are very expensive.

A drill is the power tool that most people buy first, and power drills have become so inexpensive that they've virtually driven hand-powered drills from the market. But if I could have only one power tool it would be a sander, because sanding is the most boring and time-consuming boatbuilding job. Power sanders speed and lighten the work, and the paper lasts longer in them than in the hand. My favorite is the 5-inch random-orbit sander, which does most jobs better and faster than a jitterbug sander. However, it sprays the dust more liberally, and for rounding corners, as when rounding a spar, the pad of a jitterbug conforms and make a better job.

Cabinetmakers like belt sanders, but they are working on flat surfaces more often than boatbuilders are, and they want a finer finish. Eighty-grit paper is as fine as a boatbuilder need use, except when sanding between coats of paint or varnish. Apart from sharpening tools, a belt sander is most useful in the boat shop for bringing the epoxy putty above fastening heads down flush with the wood. Dried epoxy is much harder than wood, especially okoume plywood, and the putty must initially stand proud of it because the epoxy shrinks in curing. If a jitterbug or random-orbit sander is used to cut it down, the wood around it is dished. Belt sanders take everything off evenly. They are heavy, awkward tools and not good in corners. They do usually come with dust bags that work pretty well, and epoxy and its fillers are not kind to the lungs.

The truly worthless tool is the "detail sander" on which a tiny triangle of paper is alleged to vibrate at supersonic speed. It's supposed to be good in corners, but no matter how long it is held there, the only wood or paint or glue that comes off is what clogs the point of the paper. You could finish wood faster with your fingernail.

A 7-inch disc sander-grinder represents just the opposite; it can take off material faster than you had in mind and gouge it, too. It spins at 4000 revolutions per minute, so the speed of the grit at the outer edge is 84 miles per hour, and the tool can do you or your work a lot of damage in a big hurry. But it can also do a lot of neat work in a hurry, and it's worth learning how to use it. Most often a sander-grinder is used on fiberglass with a stiff rubber pad (a "hardback") backing up a disc of perhaps 16-grit paper. That cuts down fiberglass overlaps in a hurry to make a smooth landing for the next layer. The chopped-off glass comes off the disc as from a machine gun.

The same tool is used with a foam-rubber pad (a "softback") and perhaps 80-grit paper for rapid finishing. It tends to leave swirl patterns behind it. I find that the hardback with finer disc of 80 grit smoothes end-grain wood better than any other tool. (Incidentally, in many lawnmower shops it is the preferred tool for sharpening blades.)

However, there are still places where hand sanding is the only possibility. In corners between chine log and plywood planking, for example, no power sander does the job. It pays to have a hardwood block about 5 1/2 inches by 3 inches by 1 inch with one long side cut at a 60-degree angle. A quarter sheet of paper wrapped around it gets the glue out sooner or later.

No matter how carefully the work is planned, sometimes some screw heads stand proud after the glue has dried and perhaps some screw points have also come out the other side. The best tool for dealing with them is a 7 1/4 -inch metal-cutting carborundum Skilsaw blade. With spacers and a couple of washers, it can be rigged to the arbor that is found in 4-inch disc sanders meant to be used in electric drills. Chucked up to the drill, the outer edge of the carborundum wheel spins at 25 miles per hour and cuts bronze screws fast—and even stainless ones fairly fast.

No matter how many clamps you have, the time comes when you wish for more. Other chapters have described light-duty shop-made clamps suited to specialized building processes: glued-lap clinker and Geodesic Airolite. But for flexibility in both the situation and the clamping force, boughten screw-clamps are best. The desirable features are depth of throat and length, but if a clamp is both deep and long it is likely to be heavy, which is not at all desirable.

The clamps that I always reach for first are sliding bar clamps; I go to C-clamps only when the bar clamps are used up. The smallest are 6 inches long; that size is often too small, and 12-inchers don't clutter up the work space too much, even when in use. It's good to have a few longer ones as well, and if you want to clamp together really distant objects you can buy clamp heads that attach to holes drilled through a 1-inch board, and then the length of the board is the limit of your clamping reach. Inevitably some of the epoxy we all work with so often gets on the clamp threads and slide bars. Where the epoxy won't pick off a metal tool readily, it can be burnt off with a propane torch, preferably outdoors because epoxy burns smelly. Polyester resin also burns off at low temperature, with even worse stink.

Everyone has a few favorite tools that are not found in all shops, and high on my list is the shoe, or four-in-hand, rasp. One face is curved, and the other is flat. One end is fine cut, and the other is coarse. It roughs out and smooths many shapes without my having to put the tool down and search for another. Like all tools, it dulls pretty quick when used on plywood, and as it can't be sharpened, it must be replaced.

The best place to put any tool down is on the floor. Put it down on the work or the sawhorse and start using another tool, and sooner or later vibration or a bump sends it to the floor anyway. Neither power nor hand tools improve from landing, especially if your shop floor is concrete. I like shallow-angle block planes, and after having several cast-iron ones break when they fell, I bought one with a pressed-steel body that is more resistant to this abuse. Humiliatingly, it is called a "high-school shop plane." Some of us are too lazy to take our own advice, but if you aren't going to put a tool away, the best place to put it is on the floor.

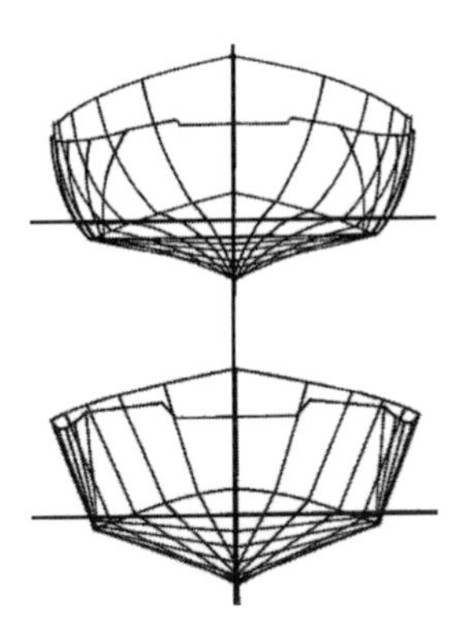

# 7

# *Phil Bolger: An Appreciation*

For very many years—perhaps since 1968, when Arthur Piver was lost in a coastal voyage aboard one of his trimarans—Philip C. Bolger has been America's most successful designer of boats for amateur builders. Piver was the editor of an insurance newspaper and knew very little about boat design in general. He claimed it didn't matter because multihulls were all that did matter, and they were best designed by inspiration. Bolger must know as much about the history and evolution of boats as any man alive, and he can make a creditable job of anything from a 12-foot kayak to a 115-foot squarerigger—and perhaps even a supertanker. But like Piver, Bolger attracts loonies, and he gives them what they come for.

Piver was notorious for his exaggerated speed claims, but Bolger's boats are not particularly fast. Whether rowed or paddled or sailed or motored, few of them win races. They express the way he feels about time spent on the water, which is not for competition but for observation and musing. In an era of insane competitiveness, his success with these designs defies explanation.

The reason Bolger wouldn't design a supertanker is that it would be too businesslike and wouldn't give enough scope to his esthetic sense. That may seem risible to some people because his simpler designs for amateurs, his "instant boats" as his sales agent Dynamite Payson calls them, are widely perceived to be boxy or ugly. Sometimes even less polite things are said about them. But these same detractors concede that when Bolger sets out to draw a traditional boat—the rowing gig *Victoria*, for example, or the daysailer *Fancy*—no other designer could make a more perfect job in every line and detail. With the instant boats the same esthetic sense is operating, but on too sophisticated a level for many people to perceive. Bolger is something like Picasso: everybody moons about his blue period, but most people whine about his later work. They do not see that his eye for beauty, better trained than theirs, has gone beyond what they are capable of appreciating.

Phil Bolger grew up in Gloucester, Massachusetts, and except for a few brief intervals he has lived there most of his life. He went to Bowdoin College, served in the occupation forces in Germany after the Second World War, and apprenticed with Lindsay Lord, the celebrated designer of planing powerboats. After that Bolger on his own picked up work wherever he could, and like most young designers and artists, some of the jobs he undertook weren't wonderful. I once knew an artist just out of school and eager for commissions who agreed to paint a mural for $50 high up on the wall of a used car dealer's office. It was to be the conventional composition: the Cadillac convertible, the blonde, and so on. But when he put the ladder up to the wall he discovered that it was rough stucco, and it took more than $50 to buy the paint to cover it. Bolger drew a good many Texas dories for Captain Jim Orell, who marketed the plans but wouldn't allow the designer's name to appear on them. "He thought he owned me," Bolger says bemusedly.

He found enough money to have a boat built for himself in 1955. Often a young designer will strain his budget to do so, in the hope of attracting new clients as well as trying out new ideas and generally entertaining himself. For example, in 1929 when Olin Stephens was a very young man, the finely built sailing yachts of old Nathaniel Herreshoff were thought too fragile for offshore sailing. What were needed were adaptations of heavy, clumsy, unweatherly commercial vessels, such as William Atkin designed. With help from his father and his partner, the broker Drake Sparkman, Stephens had built the yawl *Dorade:* 52 feet by 10 feet, fine-ended, light, with high ballast ratio (and of course with Dorade vents). In her he won the 1931 Transatlantic Race by two days boat-for-boat, and then the Fastnet Race; a ticker-tape parade up New York's Broadway followed. Stephens has lived to such an incredible age and has always worked within the racing rules that he later helped write, so he has had a very conservative influence on yacht design, and it is hard to remember what an innovator he once was.

Bolger never cared much for racing rules, and in 1955 he had no rich backers to tap, so his *Blacksnake* was more modest than *Dorade* but even more innovative. She was a double-ended sharpie sloop 33 feet by 6 feet, 2 inches, with incredibly low freeboard and a cabin that had only 2 feet of headroom but otherwise included many of the features still found on his sailboats: unstayed rig with vertically cut sails and no battens, balanced club jib, centerboard, extremely shallow rudder. "As far as I'm concerned, Captain Nat Herreshoff wasted his time when he devised cross-cut sails," Bolger wrote at the time.

In fact, cross-cut sails disturb the air flow over the cloth less than vertical seams, so the boat goes faster. Is that wasting time? And although battens often broke or slipped out of their pockets when Bolger was young, they have been improved since and are now as trouble-free as any piece of sailboat gear. Bolger admitted that a balanced jib combined with an unstayed mast drives a boat to windward only in light air and in heavy air drives it to leeward. The shallow rudder took some getting used to, he says, but nowadays he puts endplates on these rudders and says they work great. The centerboard and its trunk take up half the space in *Blacksnake*'s cabin and cockpit, but his fear of grounding daggerboards is so great that I know of only one design where he uses them (the 31-foot folding schooner).

In general Bolger's sailboat designs do not attract the keenest sailors. They tend to be lightly canvassed because they have odd ballasting arrangements (such as water, or in the case of *Blacksnake* a wooden bottom 4 inches thick), which don't work as well as a slug of lead deep down in the water. The rigs tend to be inefficient for their area because they are unstayed or because they are lugs or gaffs or some other four-sided anomaly, even in very large sizes. Sometimes they have no boom, and Bolger puts two or three masts even on very small boats. Bolger has a great deal of sailing experience, but one senses that he isn't all that easy with sailing and doesn't much enjoy it. Like L. Francis Herreshoff, when Bolger dreamed up an ideal sailboat race, it was one in which auxiliary motors could be used. But as everyone concedes, sailboats are the most beautiful boats, and Bolger knows that, too.

In fact, the whole idea of *Blacksnake* was looks. He was fascinated by *Bonnie Lassie,* an L. Francis Herreshoff R-boat, and *Blacksnake* was intended to look like her, even if he didn't have the money to make her sail like *Lassie.* "For a cheap boat she looks pretty good and gives a pretty good sail," he concluded. As sailing even a fast sailboat is likely to be the slowest way to get around on the water, one wonders whether a pretty good sail is good enough. What about a pretty good row or a pretty good paddle? What would Olin Stephens make of this?

Bolger apprenticed with a powerboat designer, and I believe that it is here and in muscle-powered boats that he has done his best work. Lindsay Lord wrote a landmark book, *Naval Architecture of Planing Hulls,* which I have tried to read more than once. It is pleasantly and clearly written and no more technical than it need be to explain the subject. But the subject itself, planing powerboats in rough water, is just so silly that I can never get through three hundred pages about them. Lord devotes quite a little discussion (as an example of these boats' utility) to the PT boats of the various factions in World War II; but then, isn't war even more pointless than the boats used in it?

No doubt Bolger disagrees. Although he is not the least combative in person and admits that he discovered while in the army that his "talent wasn't for soldiering," he is very interested in guns (he owns "an arsenal") and thinks that everybody should have as many as he likes. He is also interested in airplanes and automobiles and all other murderous instruments. Probably he could quote a good deal of Lord's book verbatim, but he is by no means a blind disciple. For example, Lord several times speaks well of sea sleds, a kind of inverted V-hull that traps air to lift the hull and lessen wetted surface, somewhat like a tri-hull (Bolger says that Ray Hunt devised the tri-hull to avoid infringing the sea sled patent). Recently there was a revival of interest in sea sleds with articles in *Boatbuilder* and *WoodenBoat* magazines. They argued with some cogency that modern materials, such as aluminum or plywood or fiberglass, might solve the structural problems inherent in the inverted V shape, and naval architect Dave Gerr thought it might be "the best high-speed hull ever."

In a letter to *Boatbuilder,* Bolger replied, "I've had hands-on experience with half a dozen sea sleds, including five Hickmans from 13 to 75 feet. I did open-water trials of two small ones and witnessed the heavy-weather trials of a 3,000-horsepower 50-footer.

"It is my considered opinion that they have no advantages over a flat-bottomed garvey of good proportions. If they are any less hard-riding, the difference is not discernable from the helm. Disadvantages compared with the flat-bottomed garvey are: poor space efficiency, reduced load capacity and/or deeper draft, complex and vulnerable construction, poor maneuverability throughout their speed range, and frequent propeller air ingestion. They are wet because they blow their spray straight ahead and high." Well, after that there's nothing more to be said about sea sleds, is there?

Art Turner, a friend and a long-time powerboat salesman, has told me that he finds Bolger's early powerboat designs "some of the loveliest I ever saw." To me, loveliness in a powerboat comes from a few simple principles, all of which make the boat less useful: great length, low freeboard, little or no cabin. The 38-foot *Sea Hawk,* which Bolger designed for the Australian Graham Hawkes, comes close to these principles, and the straight sheer in no way detracts from her beauty. Bolger worried about the difficulty of planking such a hull shape, but if he had known Hawkes he wouldn't have. Hawkes is as truly a master builder as anyone living today, as we found out when we visited his small yard in Queensland. He still has access to virgin-forest timber, and he selects and uses it with perfect unstudied ease. In *Sea Hawk* he takes tourists around to sightsee and dive on coral. We had a little correspondence later because, like Bolger, Hawkes is preoccupied with the looks of his boats and he thought that in another one he might like a round stern and perhaps a sail or two, just for theater really.

*Shivaree,* the outboard powerboat that Bolger currently owns, is too high-sided to look beautiful to me, but it does seem to work very well. Four years ago, after Carol and I had visited Bolger and his wife, Susanne took this 16-footer out Annisquam Canal behind our *Dandy* and into Gloucester Harbor. While we sailed first with working sail and then with spinnaker, Susanne circled us, taking pictures. *Shivaree* seemed supremely manageable, and at a variety of speeds and courses left Susanne free to concentrate on her camera.

Although Bolger doesn't build boats, he is interested and knowledgeable about building processes and always eager to try something new. Living as he did for many years aboard his motorsailer at David Montgomery's boat yard, he found in Montgomery a genial companion ready to try anything he dreamed up. *Shivaree* was to be carvel planked over molds without any framing worth mentioning. The fastening would be 5200 polyurethane caulk, put in at the same time as the planks. Whether this would have worked or not Bolger still doesn't know, and my own opinion is that although 5200 is tenacious, it is too flexible and the hull would have lost some shape at least when the molds came out. In any event, when the hull was about one third planked Bolger looked at the polyurethane bill and decided that steamed frames and screws wouldn't be so bad after all.

Among his muscle-powered boats the *Kotick* kayak seems a very good and useful model, although I don't know how many sets of plans have been sold for it. At 15 feet she is shorter than the gung-ho kayakers like their boats to be and more

Phil Bolger (left) with Tom and Carol Jones in *Puxe*
(Photo courtesy Susanne Altenburger)

suited to poking around than to sprinting. She is very low sided and without much crown to her deck, and to get enough space for feet and knees she has a kind of superstructure that Bolger calls a "cockpit trunk," making her look quite different from most designs, which get the needed height with more deck crown. She has more freeboard in bow than stern and is also asymmetrical below the waterline.

She was strip-planked by Dynamite Payson of 1/4-inch-by-3/4-inch strips, and constructing her can't have been too different from building doll's-house furniture or a model. This strip planking with thin strips has been a popular way to build canoes and kayaks for a long time now, but usually it is done with white glue, which is not waterproof but sands as readily as pine or cedar strips. The strips are just a core material, which is then covered with epoxy glass outside and in. Glass brings its own finishing problems, but Payson built the *Kotick* with waterproof glue and without sheathing. Such glue is far harder than softwood strips, which sand away and leave the glue standing proud, so finishing this boat—and she is very fair—must have been a nightmare of long boards and other tricks.

From the start Bolger was concerned with the hull's fragility, aware that the price Payson had charged him to build her was a bargain not likely to be repeated. He soon had a mold pulled from her so that she could at least be duplicated in fiberglass, though a plastic *Kotick* wouldn't weigh the 40 pounds that the wooden one originally did. Some years ago a combination of the boat's increasing age and

Bolger's decreasing nerve caused him to have her fiberglassed on the outside. She still looks nice, and he still uses her, but she doesn't still weigh 40 pounds.

To me the loveliest and most useful of Phil Bolger's designs (although he has designed more than six hundred boats, and I certainly don't know all of them) is the 15-foot, 4-inch by 4-foot, 6-inch rowboat *Spur II*. "She was design number 400, and I wanted to draw something special." Like *Kotick, Spur II* is not outlandish or extreme in any way and not a boat for the racing crowd. David Montgomery built her of 1/4-inch glued-lap okoume ply, and her only frames are forward and aft of the three thwarts plus one more near the bow. These are of 1/4-inch plywood also, and all weights in her are watched very carefully to arrive at a boat of just under 100 pounds.

*Spur II* could best be likened to a Whitehall, although I never saw a Whitehall as pretty or as useful. One hundred pounds is some weight to carry, but not too much to slither up onto a floating dock or to turn over. When Susanne rows her solo (and Susanne is muscular), she literally spurts through the water. Put two more people aboard, and they slow her hardly at all. Put another 500 pounds in her, and she would probably still row respectably. This boat is almost miraculous to me in her simplicity and elegance and the functional perfection of every detail. Nor is she difficult to build, apart from the tedium of glued-lap planking. She has no hollows in her sections, as do traditional Whitehalls.

And yet, what is the market for such a boat? She is not a sliding-seat exercise machine. She is too big to be a tender to any but a very large yacht, which would more likely carry a planing powerboat in davits, somewhere near the helicopter pad. Perhaps she is nothing more than just the best rowboat anyone ever designed. Feeling the limited market for her, Bolger drew her a rather bad sailing rig, which he included apologetically in a book chapter about her.

Bolger has designed many other excellent rowboats, which look less than perfect only when compared to *Spur II*. The already-mentioned *Victoria,* designed twenty years earlier, is another glued-lap boat, only 2 inches longer and 4 inches narrower. He specified a full inch overlap for the planks, not then trusting these joins as much as we have all come to do since; and that combined with huge stern and bow sheets, greater freeboard, and so on, make her a heavier boat. More troublingly, *Victoria* has a greater bow overhang and a somewhat raked transom, making her waterline a good foot shorter. Her waterlines are hollow forward, which I do not think an advantage, though some might argue the point; and while *Spur II* has eight planks to a side, *Victoria* has eleven and hollow sections forward, so she would certainly be more work to build.

As Bolger says in *Small Boats, Victoria* was derived from a pulling boat that he designed and had built while working for Stanley Woodward in Mallorca. The Spanish boat was called *Spur* and was designed for looks above everything. She had a great deal of sheer and a black hull with a gold sheer stripe and a gold centaur plastered on the transom. In a photo I saw she has a 6-square-foot American flag floating in the breeze, which must have added to the pleasure of rowing to windward. Bolger is at the oars, and in the stern sheets sits a blonde who is also of considerable decorative value. Bolger says that an article about *Victoria* in a boating magazine produced about 1,500 plans inquiries. Using the normal scaling factor of 1:20, that would have resulted in 75 plans sales, which is pretty good for a rowboat. Twenty years later I can't imagine that *Spur II,* which I think a somewhat better boat, sold half as many.

Another kind of rowboat that Bolger has spent many years perfecting is the "light dory." The best-known dories were developed for the Grand Banks fishery, and they had to be inexpensive to build, stackable on a schooner's deck, capable of carrying a big load of fish, and seaworthy to some extent. They didn't have to be especially good rowing boats, and they weren't. The sequence of lower, lighter, plywood dories that Bolger developed rowed better. He drew one for Captain Jim Orell, the Gloucester Gull, and the most recent version that I know of was designed to Bolger's beloved metric system. The tombstone transom has gone, and she has become a pure double-ender. To simplify construction he has put the chine log on the outside of the hull, and he says in *Small Boats,* "I don't think it increases resistance, but I can't prove it yet." I would think that a little tow testing even with a simple balancing wand in the current of a stream would soon put paid to that idea.

One sees an enormous number of Bolger dories, especially in New England. If there is a wooden boat amongst the inflatables and plastic bathtubs at a dinghy dock, it is usually a Bolger dory. Inevitably people tow them behind their cruising boats, which does nothing for cruising speed; but equally inevitably people tow inflatables, often with the outboard leg dragging in the water, and I'm sure a light dory generates less drag than that. I've heard no complaints about their shipping water in a normal seaway. People like their light dories, which aren't as good rowboats as *Spur II,* of course, but are lighter and somewhat cheaper and vastly easier to build. Bolger says that the Gloucester Gull will be his ticket through the Pearly Gates, and I think that as long as some cruising yachtsmen have the sense to row when they reach port and get their blood circulating again, we are likely to see these dories around.

Bolger's writing is another expression of his enormous energy and industry. Most boat enthusiasts know that he has written five books describing his designs and one about sailing rigs. Most also know that he has written innumerable magazine articles, starting early in his career (he was writing about *Blacksnake* in 1955). But Bolger estimates that he also writes a thousand letters a year, and he has written a number of novels, too. It is hard to see how he could have accomplished all that he has had he not waited to marry until relatively late in life.

A thousand letters a year is nearly three a day, and all that I have received have been nearly as succinct and felicitously phrased as his books and articles. They are not brief—always a full page, sometimes more—and they are as likely to be handwritten as typed. On the other hand, none of his novels has found a commercial publisher. I have only read one, *Schorpioen,* which was published by the boatbuilders Duff and Duff. The trouble with the book, and with all Bolger's novels judging from his descriptions of them, is that they are about political and economic ideas, not about people.

Talking with Phil Bolger, I am sometimes reminded of Ezra Pound, whom I went to visit when I was in college and he was confined to St. Elizabeth's Hospital in Washington, D.C. I had looked forward to an afternoon of literary insights, but what he gave was a lecture on politics and economics. Of course, Pound was nuts, and Bolger certainly is not; but when we're with him we have to work hard to keep the conversation on boats or people or nature. Foolishly, I once said to him that Captain James Cook was the greatest man who ever lived. He replied that Cook had to come second to Adam Smith. Smith was an obscure academic who published two books, one of which, *The Wealth of Nations,* has a tenuous connection with Bolger's beloved Libertarian politics. There isn't room here to describe Cook's humanity or list his accomplishments, but to hear him compared to Smith is like listening to Uncle Ezra all over again.

Bolger's writing about boats is another matter. As I have said elsewhere, he is the best writer who ever wrote about boats. Like most authors he didn't write masterpieces from the start: his articles on *Blacksnake* are well worth reading but a little diffuse. However, long before the publication in 1973 of *Small Boats,* his first book of designs, he had mastered the writing craft. Many of us who had not kept up with all the boating magazines and were not aware of Bolger's ascending star had a genuine shock of pleasure from this book and went back to reread it only weeks after we had finished it the first time. We were weary of Billy Atkin, with his clunky boats and his ersatz salty talk. We were fed up with the prissy L. Francis Herreshoff. We were revolted by the Corinthian condescension of Francis Kinney. Here was a designer with only the most minor axes to grind (he devotes most of the preface to attacking the "foot-inch English measurement system," and predicts its demise, prematurely as it turns out).

The boats in Bolger's first book are all under 28 feet, and most are under 20 feet, but what a variety of styles and propulsion and materials and building techniques! And Bolger knows the trade-offs: the ease or difficulty of planking certain shapes and the resistance likely to result; the pleasure and labor of building, of owning, of using; the suitability of materials and the compromises in building that we are all tempted to make. Each chapter describes one boat, and in each there is some narrative, some design discussion, and some talk of building and use. Perhaps the format was originally devised by Atkin for his interminable series in *Motorboating,* but reading Bolger is like listening to a string quartet and reading Atkin is like hearing a cowbell.

Who could pick out a favorite among Bolger's design books? They all have a great many good boats and good jokes and interesting ideas, and whichever one of them was published first would have given us the same shock of pleasure that *Small Boats* did. Perhaps taking a cue from that title, in 1978 Roger Taylor began publishing *Small Boat Journal.* It was a heaven-sent opportunity for Bolger and for all of us who were writing about boats at that time, and considering how much money the magazine lost, it is remarkable how long it lasted.

Bolger came aboard with volume 1, number 1, and I believe he wrote a piece for every issue until 1989, when *SBJ* was sold south and lost its name and identity in a haze of outboard fumes. His contribution evolved into the "Bolger Cartoon," each one starting with a letter from a reader who wanted a design for a certain type of boat not then on the market. Bolger would make a sketch or study plan and comment on it, much as he does in letters to prospective clients, except that in the magazine the inquirer was less likely to become a client and he could answer with less reserve. To a man who wanted to take a dog along on a retirement houseboat, he replied, "The dog is your problem. I suggest that a good-sized python would be better adapted to life afloat . . ." Much of the contents of Bolger's later *Boats with an Open Mind* consists of reprints of these SBJ articles, and it's good to see them again.

Bolger's writing seems to flow forth to us as naturally as conversation, although in fact it is the product of labor, and Bolger is not an especially easy or well-organized conversationalist. He does say interesting and funny things when he speaks as well as when he writes, but he is not fast on his feet, and one feels that his shrewdest and best-phrased observations are something he has thought of before and perhaps said in a book or article. He uses remarkably few commas in his writing, but the wording is so carefully thought out and logically arranged that one never pauses to wonder what he means.

To my mind Bolger's best piece of writing is an article not about his own designs but about those of L. Francis Herreshoff that appeared in the now-defunct *Nautical Quarterly* ten years ago, before Carol and I were friends with him. I wrote him then to say how much I liked the piece and that I hoped he would write about other designers and perhaps accumulate a book. He replied that there simply were no other designers of equal interest. Let us see what Bolger found unique in the idiosyncratic and fusty Herreshoff.

The first half of the article is devoted to the highly engineered racing boats that consumed the first half of Herreshoff's designing career. A whole drawing might be devoted to one turnbuckle so perfectly proportioned, says Bolger, that if stressed beyond its limit it would "vanish in a fine molecular mist like the proverbial one-hoss shay." But Bolger thinks that L. Francis made a greater contribution than his father Captain Nat, "speeding the pace of a minor technology like yacht design."

The second half concerns cruising designs from the second half of Herreshoff's career. Bolger thinks Herreshoff could "conceive the complete boat in all its details and in three dimensions in one flash, as a large room, full of furniture or machinery, is illuminated by switching on a light." Furthermore, the 58-foot ketch *Bounty* is "the most beautiful yacht ever designed or built, a flamboyant beauty with a riot of sweeping, twisting, converging curves, all set off with intricate details and ornament, all blending into a perfect traditional overall effect."

Bolger says, "Herreshoff considered himself an artist primarily." He compares Herreshoff to Leonardo da Vinci and says that he was "producing toys in about the same sense that the *Last Supper* is wallpaper." Although he puts it eloquently, Bolger is certainly wrong about this. It may be that da Vinci's patron specified the subject of his painting, and we know that the subjects in Rembrandt's group portraits all had to be the same size because they were all paying the same amount. But the crafts—and the designing of boats is a craft, not an art—are much more confined, because a boat must float on her marks and move through the water, just as a pitcher must pour predictably and a barber's haircut must look like the one that you pointed out in the magazine. What's wrong with Herreshoff's cruising designs is that function is sacrificed to looks. Bolger himself doesn't make that mistake unless pushed hard by a client.

Of Bolger's recent work, most people think of the "instant boats," although he probably spends more of his time on larger one-off designs. When we visit he always has several in the works, and many are motorsailers, flat-bottomed with box keel, made of plywood or steel. They often look like giant instant boats. For someone like Stanley Woodward he can draw a more traditional vessel, but increasingly his penchant is for high-sided, flat-bottomed models. Even his round-bottomed sailboats usually have a tight bilge radius, with a flat bottom and plumb topsides.

He has a good many calls for liveaboard designs, partly because he lived aboard himself until quite recently, and the newer ones tend to be shorter and stockier than his own lovely *Resolution.* From years of experience he has discovered how seldom such a boat is under way, and he feels that the greater speed of the longer, narrower model does not justify the greater cost of building and maintenance to arrive at the same accommodation. He prefers copper sheathing to bottom paint, especially for a liveaboard boat, and says that he gets nearly twice the life from it as did long-time cruiser Eric Hiscock, who was under way so much.

The smaller designs, intended for a high volume of sales and perhaps not drawn for a specific customer in the first place, are priced very inexpensively. Often sold by middle men, such as Dynamite Payson or Common Sense Designs, these sales bring Bolger an income without a deluge of after-market questions. However, his popularity means that he cannot have a listed telephone, and lately he has been getting as much grief from the phone company (apparently it is bad form to have an unlisted commercial phone) as he would from his builders.

The first thing to be said about the "instant boats" is that all the owners I have talked to like them. Al Whitehead's satisfaction with his Bobcat has already been described in chapter 2. Another Bobcat we encountered in Mystic was very much cruder in execution, having been built by a novice and tape seamed. But the owner was pleased with his boat and pleased with his accomplishment. So it was with the Cartopper, a 12-foot rowboat that was tied astern of a fiberglass cruising sailboat in Block Island. The skipper said the family would soon be arriving to use the cruising boat as a summer cottage, and their total weight was too much for a light dory. The taped seam inside was so rough that I'd have worn gloves before putting my hand on it, but many a fiberglass boat is little better. The skipper was pleased with his Cartopper and hoped it would last forever, because he wasn't eager to build again.

In the Kickamuit River, Rhode Island, where we were weather-bound for 24 hours last summer, a bright orange June Bug wove back and forth through the moorings, generally using oars downwind and an outboard upwind. The owner said he had built her in a couple of weekends, although I think that an exaggeration as he had never before built a boat. He had the sailing rig for her, but he rowed and motored her mostly, as he found sailing a little tricky. He liked her a lot, and in fact, the only owners I've met who have been disappointed with their instant boats have been those who built them of bad materials, using underlayment instead of marine plywood, for example, or polyester instead of epoxy resin, and had them fall apart.

The June Bug is exactly the kind of instant boat that pulls so many people's chain. She is 14 feet by 3 feet, 3 inches and perfectly plumb-sided and plumb-ended. As a result, the owner mounted the outboard (which Bolger does not suggest for her) without thought to the 15-degree transom rake that is normal for outboards, so probably it pushed the stern down a bit. But even with three good-sized men in it—perhaps a total load of 700 pounds—the 3-horsepower motor pushed the boat to hull speed with very little effort; and no microscopic alignment of propeller angle and waterline would have pushed it any faster.

That June Bug probably never has been and never will be outside of the Kickamuit River, 1/2 mile wide and less than 2 miles long. She didn't look like other boats, but she did look like fun. Writing about her in his fourth design book, *30-Odd Boats,* Bolger defends her against the "purists," who would compromise her utility—her large carrying capacity combined with small overall dimensions—for the sake of some preconceived esthetic standard. Her sheer line is artfully drawn, and in the prototype that Payson built it is emphasized by two-color topsides. The effect is jaunty and pleasing, and we liked the Rhode Island boat, too.

Bolger did not always design small boats for "instant" construction. Since *Small Boat Journal* died, Bolger has been writing for *Messing about in Boats,* a counterculture magazine. In a retrospective article there recently, he discussed a 20-foot daysailer that he designed for his brother even before *Blacksnake.* He said, "My brother, who built six or eight boats for himself over a forty-year period, took it for granted that lofting, building a ladder frame, and torturing some panels was part of building a boat. I've spent a sizeable part of my working life trying to speed up the process for people who like sailing more than carpentry, who want a boat as soon as possible." Of course, he is not against people who like both carpentry and sailing, and for them he has *Spur II* or even the 115-foot squarerigger. If asked what he thought of people who built one instant boat after another and eventually had a whole garage full and indirectly laid acres of forest to waste, he'd probably give some stock libertarian answer and point out that each plan sold brought him a royalty.

Bolger says that his one genuinely original contribution to yacht design is his Birdwatcher, which he has now drawn in various sizes. These sailboats have high cabin sides running nearly the whole length, which are almost entirely Plexiglas, and the cabintop is slotted to give standing headroom from bow to stern. Apparently these boats can recover from a knockdown, and they give "comfort as well as safety," he says. Another contribution of his, certainly original, is the break-down nesting tender with joins running fore-and-aft rather than athwartships. For several years he experimented with a daysailer that had its rudder forward and daggerboard aft. And what of the folding schooners? Bolger certainly underestimates his originality.

He loves to fiddle with plywood shapes, to see how economically they can be fitted onto 4-foot-by-8-foot sheets. He loves designing and drafting in general and takes pride in his work. Unlike the artistic L. Francis Herreshoff, Bolger's plans have always been easily read and attractively laid out. Recently at a symposium that featured him and the 92-year-old Olin Stephens, Bolger was eager to hear which of the early designs Stephens actually drew himself. Stephens replied that from the time he could afford to hire a draftsman, he never drew a boat himself. Bolger was chagrined. Nonetheless, he expects to go on drafting as long as he works. On the other hand, Susanne is many years younger than he is and wants to design too, and he feels that for her computers are the only sensible choice. They have spent some thousands of dollars on software, and it would be a mistake to think that a hundred dollars of this stuff would make one a naval architect. One must also learn how to use it.

Bolger has thrown himself entirely into his marriage with Susanne Altenburger. Nearly everything leaving the office now, whether drawn or written, says "Phil Bolger and Friends," meaning the two of them and her seven cats. She is a big, loud, forceful German woman. The first time we met her, the tide was out at Montgomery's yard, and we had to anchor in the Annisquam River and wait several hours to come ashore. Susanne in huge gumboots came mushing down to the shore, full of largely unintelligible shouts, pantomime, and good cheer. The time passed quickly. We have heard that she can be abrasive and even Bolger mentions it, but to Carol and me she has always been a warm, funny, uncompetitive friend. She came to him originally to learn about boat design, but after she "tried it out," as Bolger puts it, they got married.

Susanne is especially interested in the mechanical systems of boats and in using off-the-shelf parts—perhaps from an auto parts store—rather than having marine parts custom made. This approach is exactly the opposite of that of L. Francis Herreshoff, who enjoyed designing every part of a turnbuckle from scratch. It requires compromises: on the electric launch *Lily,* which is largely her design, the batteries are trimmed forward and aft to allow for crew loading. It is done with a crank turning a threaded rod—a stock item—and moving the batteries about 1 inch for every ten revolutions. The batteries, on tracks and rollers, could easily be moved that distance with a single revolution. Susanne is so vigorous that she'd as soon turn the crank two hundred times as twenty. But other people may want something different.

In the latest designs it is hard to tell which elements are her ideas and which are his. She favors the flat bottom and box keel shape, but he was designing such hulls (for example, a 16-foot steel tug) long before they met. As time goes on, the distinction is likely to become more and more blurred. But Bolger seems bright and focused. After the memorial service for his brother, he said, "In recent years I've sensed that my brother was losing interest in things. I was beginning to lose interest myself, until Susanne came along."

# *Designers' Addresses*

Below are listed the addresses of designers whose boats are discussed in some detail in this book. They and other designers will be more likely to respond when a stamped, self-addressed envelope accompanies your inquiry.

Phil Bolger and Friends, Box 1209, Gloucester, MA 01930

Joseph C. Dobler has died. Redrawn plans for his 16-foot plywood skiff are available from Jones Boats.

Jones Boats, Box 391, Tuckahoe, NJ 08250

Monfort Associates, 50 Haskell Road, Westport, ME 04578

Robert W. Stephens, PO Box 166, Brooklin, ME 04616

# Index